The Physics of History
Part II

Professor David J. Helfand

THE TEACHING COMPANY ®

PUBLISHED BY:

THE TEACHING COMPANY
4840 Westfields Boulevard, Suite 500
Chantilly, Virginia 20151-2299
1-800-TEACH-12
Fax—703-378-3819
www.teach12.com

Copyright © The Teaching Company, 2009

Printed in the United States of America

This book is in copyright. All rights reserved.

Without limiting the rights under copyright reserved above,
no part of this publication may be reproduced, stored in
or introduced into a retrieval system, or transmitted,
in any form, or by any means
(electronic, mechanical, photocopying, recording, or otherwise),
without the prior written permission of
The Teaching Company.

ISBN 1-59803-546-0

David J. Helfand, Ph.D.

Professor of Astronomy, Columbia University

Professor David J. Helfand is Professor of Astronomy at Columbia University. Following his public high school education in Mattapoisett, Massachusetts, Professor Helfand attended Amherst College as a scholarship student from 1968 to 1973, the extra year being occasioned by his draft lottery number of 12 and an Independent Scholars program that allowed him to begin research at the University of Massachusetts. He received his Ph.D. four years later while working under Professor Joseph Taylor, the 1993 Nobel laureate in Physics.

Professor Helfand moved to Columbia University in 1977 to accept a position as a postdoctoral fellow in the group under Professor R. Novick, which was preparing for the launch of the first X-ray telescope: the Einstein Observatory. In 1978 he was appointed Assistant Professor of Astronomy, and in 1982, Associate Professor of Physics. Following a year at the Danish Space Research Institute in Copenhagen, Professor Helfand took a 10-year sojourn as a faculty member in Columbia's Department of Physics before returning to the Department of Astronomy in 1992.

Owing to his longstanding opposition to the tenure system, Professor Helfand does not hold the usual "appointment without term." Instead, he has served under a series of five-year contracts that require a formal evaluation of teaching, service, and research, followed by a provostal decision for termination or reappointment. As of 2009, he is in the third year of his sixth such contract. After a year as the Sackler Distinguished Visiting Astronomer at the University of Cambridge, Professor Helfand is again a Professor of Astronomy at Columbia University. He served as the department chair at Columbia for more than 10 years, until he was liberated in 1997. Unfortunately, he was recaptured in 2003 and again serves in that capacity.

Professor Helfand's work has covered many areas of modern astrophysics, including radio, optical, and X-ray observations of celestial sources ranging from nearby stars to the most distant quasars. He is currently involved in a major project to survey our galaxy with a sensitivity and angular resolution a hundred times greater than is currently available; the goal is to obtain a complete picture of the birth and death of stars in the Milky Way.

Professor Helfand primarily teaches undergraduate courses for nonscience majors, including one of his own design that treats the atom as a tool for revealing the quantitative history of everything—from the human diet and works of art to Earth's climate and the universe. This is the course he shares with us as *The Physics of History*. He also designed the Frontiers of Science course that Columbia added to its famed Core Curriculum in 2004 as a requirement for all freshmen, the culmination of a vision he has pursued since 1982.

Professor Helfand received Columbia's 2001 Presidential Teaching Award and the 2002 Great Teacher Award from the Society of Columbia Graduates. During the late 1990s, he appeared weekly on the Discovery Channel program *Science News*, bringing the latest astronomical discoveries to the U.S. television audience. More recently, his television appearances have included more serious matters on Comedy Central's *The Daily Show* and the National Geographic Channel's *The Known Universe*. In 2009, he took a leave from Columbia to serve as the president of Quest University Canada, a new liberal arts and sciences university he helped launch in 2007 to foster intensive, interdisciplinary undergraduate teaching and learning.

Professor Helfand serves on far too many university, government, and American Astronomical Society committees for his own (or anyone else's) good. He believes he is a better cook than astronomer and, ambiguously, most of his colleagues who have sampled his gastronomical undertakings agree.

Table of Contents
The Physics of History
Part II

Professor Biography i
Course Scope 1

Lecture Thirteen	A Bad Day in June—Death of the Dinosaurs	 4
Lecture Fourteen	The Origin and Early History of Life	 18
Lecture Fifteen	The History of Earth's Atmosphere	 34
Lecture Sixteen	The Age of the Solar System	 50
Lecture Seventeen	What Happened before the Sun Was Born?	 65
Lecture Eighteen	Atoms Are Star Stuff— Cooking Up Carbon	 79
Lecture Nineteen	The Lives of Big Stars— Cooking Up Big Atoms	 96
Lecture Twenty	Relativity— Space and Time Become Spacetime	 112
Lecture Twenty-One	(Almost) Everything Is Relative	 126
Lecture Twenty-Two	Matter Vanishes; Light Speed Is Breached?	 142
Lecture Twenty-Three	The Limits of Vision— 13.7 Billion Years Ago	 156
Lecture Twenty-Four	The First Few Minutes— Where It All Began	 172

Timelines 188
Glossary 195
Biographical Notes 212
Bibliography 217

The Physics of History

Scope:

The history of the universe and all it contains is written in the particular arrangements of the fundamental particles that make up atoms. Given an understanding of the physical laws describing the behavior of these particles and the energies with which they interact, we can "read" that history, just as, given a vocabulary and the rules of syntax, we can read the record of human history, consciously recorded.

Atoms are not culturally biased like human historians, although heavy ones move more slowly than light ones, and we will need to remain alert to such effects as we proceed along with our atomic guides. More importantly, however, atoms provide us with access to times far earlier than those human historians have recorded, allowing us to explore the rise and fall of preliterate societies, the history of Earth's climate and the evolution of our atmosphere, the origin of the solar system, and even the origin of atoms themselves, beginning with the formation of their constituent particles in the first few microseconds of the universe.

Atoms are tiny. Arranged in a single layer, 25 trillion of them can dance on the head of a pin. Their internal structure is, indeed, an elaborate minuet of charged particles, and the rhythm of their dance produces radiation that can be used to identify them across billions of light-years of space. Remarkably, the atoms we see out there are identical—precisely—to those of which we are made.

We will begin this series by getting to know our little historians intimately—how they are constructed, how they interact, and how they are involved in transmitting to our brains everything we know about the world. Examination of isotopes—atoms with identical chemical properties but with crucial differences in mass—is central to our story and will lead us to an understanding of the imperturbable little clocks that allow us to construct calendars for times long past.

Examined on the atomic scale, art forgeries are easily detected by discovering which elements are present (and absent), all without touching the painting; illuminating the canvas with neutrons gives each type of atom a transient glow that can be identified by the time it takes to fade, its characteristic "half-life." The haunting cave paintings of our forebears reveal their secrets through analysis of their carbon-14 content, which immediately provides a date for their creation. Ancient buildings likewise give up their history under an atomic gaze; the rate at which plaster cures allows us to date a building to within a decade of its construction. And since we are, quite literally, what we eat, the atoms of ancient bones allow us not only to date our ancestors' demise but to reconstruct in stunning detail their diets and their nomadic wanderings. A history of agriculture and of human migration is opened to our view.

Trees, too, are what they eat, and an atomic analysis of their annual rings reveals the isotopes they ingested, allowing us to reconstruct both temperature and humidity over the past 12,000 years with a precision comparable to that of modern weather stations. Annual layers of glacial ice in Greenland and Antarctica contain an even more remarkable record of past climate, extending back in time nearly 800,000 years, long before our species emerged on the plains of the Serengeti. Bubbles of air trapped in the ice act as atmospheric samplers from the distant past, while embedded salt and sand tell us of wind speeds in prehuman times, and the ice itself records the temperature. Probing even further into the past, layers of sediment from the ocean floor record tiny changes in the isotopic composition of the shells of microscopic sea creatures, providing a thermometer of stunning accuracy while also recording events that changed the evolution of life on Earth, such as the death of the dinosaurs 64.5 million years ago.

Changes in Earth's atmosphere and in its orbit about the Sun are reflected in these long-term records and allow us to unravel the forces that modified our climate long before we developed the technology to do so ourselves. We are not, in fact, the first creatures to change significantly the composition of our atmosphere; the story of ancient life is written in rocks at the atomic scale, and reading it carefully allows us to envision nearly all of Earth's history. But we need not stop there. Analysis of ancient meteorites, the detritus left over from the formation of the solar system itself, tells us not only of the Sun's first years but also, remarkably, what preceded its formation in our galactic neighborhood.

Our story does not originate a mere 4.56 billion years ago when the first meteorites formed. Nearly all of our atomic scribes were created in the interiors of stars that died even longer ago, during the first 8 billion years of the Milky Way's history. It is the lives of these ancient stars that will allow us to understand the history of the atoms themselves.

Once we move beyond our galaxy, we need to examine the surprising relationship between two seemingly disconnected concepts: space and time. Einstein's deep insight into this problem a century ago has forever changed our view of reality. A primer on relativity will allow us to proceed further into history as we understand that looking out into space is looking back in time.

Our story remains incomplete until we understand how the first atoms were formed and their constituent parts created. The concluding lectures will be devoted to this quest. Precise measurements of hydrogen isotopes in the distant universe, seen through the absorption of distinct wavelengths of light by clouds of gas that have yet to form galaxies, coupled with an analysis of the light set free when the first atoms formed, allow us to trace our story back to the initial microsecond of the universe. It was then that the primeval quarks came together to form protons, the building blocks of all matter. Following one quark from its birth through the history its fellow particles have revealed might find it lodged happily in the ink dot at the end of this sentence.

Lecture Thirteen
A Bad Day in June—Death of the Dinosaurs

Scope:

Sixty-five million years ago, the Earth was ruled by dinosaurs. Sixty-four million years ago, they were all gone, setting the stage for the rise of mammals. What happened? One day in June, 64.5 million years ago, an asteroid roughly 10 kilometers in diameter crashed into the Yucatan Peninsula of what is now Mexico. Forest fires were ignited around the world within hours. Thousands of cubic kilometers of sulfur-rich rocks were vaporized and thrown into the atmosphere, darkening Earth for months. When the sulfur, after combining with water vapor, rained back to Earth as sulfuric acid, 10 kilogram of it drenched every square meter. Our first clues to this catastrophe were discovered in a layer of rock from that era that was found to be heavily enriched with the element iridium, which is very rare in the Earth's crust but abundant in asteroids. Subsequent analysis from rocks and ocean sediments throughout the world provide a detailed picture of this major event in the history of life on Earth.

Outline

I. In 1979, it had long been known that the disappearance of dinosaur fossils occurred relatively abruptly roughly 65 million years ago, where "abruptly" in geological terms means "in less than a few million years."

    A. In geologic era terms, this is known as the Cretaceous-Tertiary boundary ("K-T boundary" for short).

    B. Walter Alvarez from the University of California, Berkeley, was looking for an accurate clock to measure ocean sedimentation rates for this period in hopes of finding a clue as to why land and aquatic species underwent such an abrupt change at this time.

    C. He chose the element iridium (77 protons), specifically ^{193}Ir.
        1. Iridium is a heavy, "noble" metal like platinum that does not form compounds easily; the bulk of it settled to the core of the Earth during its formation.

2. Most of the iridium in the surface crust today comes from interplanetary space, riding in with the roughly 10,000 tons of space dust that settles on Earth each year.
   3. ^{193}Ir is not radioactive; the "clock" in this case comes from the fact that the infall rate of iridium is constant, whereas the ocean sedimentation rate is not, so the concentration of iridium is low when the sedimentation rate is high and vice versa.
   4. In a layer of sediment from Gubbio, Italy, Alvarez found that the iridium concentration jumped from 0.3 parts per billion to 7 parts per billion precisely at the K-T boundary, above which no dinosaur bones exist.
   5. After finding the same iridium anomaly at sites in Denmark and New Zealand, Alvarez published a paper in 1980 claiming a meteor impact killed the dinosaurs.
   6. Catastrophism was back, and most scientists simply didn't believe it.

II. Thirty years later, the evidence for a meteor impact and its global consequences are overwhelming.
   A. The iridium layer has been found at the K-T boundary in more than 120 sites worldwide, with a distribution of concentrations decreasing away from the Caribbean Sea.
   B. The crater made by the impact has been found.
      1. The Chicxulub crater, now filled with sediment, was found during oil exploration in the Yucatan Peninsula of Mexico.
      2. It is roughly 200 kilometers in diameter, implying an asteroid roughly 10 kilometers in diameter hitting with an energy roughly 50 times that of the world's entire 10,000-megaton arsenal of nuclear warheads.
      3. Rock was vaporized to a depth of 2 kilometers; 900 kilometers away in Haiti, the ejecta layer of Chicxulub material is 0.5 meters thick.
      4. In addition to being rich in iridium, the ejecta layers contain fused quartz, tectite glasses, diamonds, and zircons, all of which only form under extreme pressures.
      5. Zirconium ($ZrSiO_4$) incorporates uranium into its crystal structure but excludes lead; subsequent uranium-to-lead decays allow accurate dating of the event to 64.5 million years ago.

- C. Direct evidence has been found that allows us to reconstruct the immediate aftermath of the impact.
    1. Layers of sandstone created by a Caribbean-wide tsunami have been found from Alabama to Guatemala.
    2. A layer of soot from forest fires ignited by the impact is found worldwide.
    3. The anhydrite rock of the Yucatan ($CaSO_4$) is rich in sulfur and produced H_2SO_4 in the atmosphere, which formed a reflective blanket, plunging temperatures from 20°C to −5°C in days; frozen lily pad fossils indicate a sudden freeze and place the time of year as June.
    4. The subsequent acid rain (~10 kg/m^2) eroded continental rocks enough to change the strontium-87/strontium-86 ratio in the oceans, as seen in planktonic shells from this epoch.
- D. Longer-term impacts came from altered atmospheric composition.
    1. After the 100 trillion tons of dust and acid rain fell out of the atmosphere over many months, sunlight could reach the surface again.
    2. The huge amount of H_2O vapor thrown up by the explosion, in addition to the CO_2 released when limestone was vaporized, caused an enhanced greenhouse effect that raised temperatures for about 10,000 years.
    3. Some species of amphibians and birds survived the fires, the months of cold, and the years of heat, but the big winners were the mammals.

III. The K-T meteor clearly had global impact on Earth's climate for millennia and profoundly changed the balance of life on Earth. But it was not alone in modifying the composition of Earth's atmosphere. Today's atmospheric composition has been present for only about 1% of Earth's history.

Suggested Reading:

Alvarez, *T.rex and the Crater of Doom*.

Questions to Consider:
1. What are the human characteristics that incline one toward a uniformitarianist versus a catastrophic view of Earth's history?
2. What factors come into play as a scientific idea goes from a radical proposal to near-universal acceptance?

Lecture Thirteen—Transcript
A Bad Day in June—Death of the Dinosaurs

Throughout history, especially pre-scientific history, catastrophism has always been a more popular notion as an explanation for how the world has evolved than slow, gradual evolution. Many religious accounts are replete with sudden and/or catastrophic events that change the world, from the creation myths prominent in all sorts of religious traditions, to Noah's Flood, to the predicted Rapture. Indeed, it was not until the middle of the 19th century that our modern view of gradual evolution came to dominate our discussion. The geological processes of volcanism and plate tectonics that shape the Earth and the biological processes of mutation and natural selection that shape the species are very slow processes, evolving and unfolding over millions of years. Nonetheless, catastrophes retain a strong emotional appeal. We like dramatic events. And now we know that they sometimes do occur on Earth, wreaking change on a planetary scale.

In 1979, it had long been known that the disappearance of the dinosaur fossils occurred relatively abruptly, about 60-some million years ago, where "abruptly" in geological terms means in less than a few million years, which is about as well as one can date the layers of fossils laid down in rock. In geological eras, this is known as the Cretaceous-Tertiary boundary, or for the German spelling of Cretaceous with a K, the K-T boundary for short. It marks a distinct change between fossils below this level, longer ago into the past, and fossils above this level. Indeed, it appears that this marks one of the points of a very large number of extinctions occurring at a specific point in the history of the Earth.

Walter Alvarez, in 1979, was a young geologist at Berkeley, and he was looking for an accurate clock to measure the rate of ocean sedimentation for this particular period of time, hoping to find a clue as to why both land and ocean species underwent such an abrupt change at this time. Indeed, by looking at all of the fossil record, it appeared that more than 50% of all species on the Earth—plants, animals, ocean creatures—went extinct at the same time the dinosaurs did. It wasn't just big creatures; it was the tiny little forams in the ocean—species of those went extinct at the same time. Indeed, some records suggest that up to 95% of these ocean-dwelling species on the low end of the food chain went extinct about this period.

Alvarez chose the element iridium as his clock. Iridium is number 77 in the periodic table, and its atomic mass is number 193. Indeed, it's one of the most massive nonradioactive elements. It's called a noble metal in analogy with the noble gases. Argon, helium, krypton don't form any molecular bonds at all because of the complete shells of electrons in their outermost orbit. The noble metals aren't quite that way, but they are also very reluctant to form compounds with other things. Platinum is the most common of the noble metals and iridium is a rare one that he thought would be useful as a clock.

The bulk of iridium settled to the core of the Earth during its formation. Just like when you pour oil and vinegar together to make salad dressing, you see the oil, which is less dense, float to the top, and the vinegar settle to the bottom. Likewise, the mix of elements that made up the Earth, which is not representative of the mix of elements in the universe as a whole, nonetheless have also separated such that the elements on the surface of the Earth are very different from the elements inside. It's instructive to actually take a look at the elemental composition of the Sun and the crust of the Earth and compare the two. The Sun is composed 98% of hydrogen and helium gas. The next most common element is oxygen, and it makes up slightly less than 1% of the Sun. The Sun is representative of the composition of the universe. Most of the universe is hydrogen and helium, and all the other elements are just traces.

The Earth, as I mentioned earlier, lost most of its hydrogen and helium very early in its phase. Helium is extremely rare on Earth today and is only there as a consequence of radioactive decay, whereas the hydrogen has escaped completely, except for that that has combined with oxygen and made water and various other chemicals on the surface of the Earth. The Earth's surface composition is, in fact, dominated by oxygen; 46.7% of the Earth's surface is the atom oxygen, 27% is silicon—silicon dioxide is sand— and the next most abundant element is aluminum, which makes up 8% of the crust of the Earth, with 5% in iron, 3.5 percent in calcium, and 2% in a number of other elements like sodium, magnesium, and potassium.

These elements are greatly overrepresented on the crust of the Earth. However, they do not represent the bulk composition of the Earth. This is obvious if one just looks at the density of the Earth as a whole versus the density of its surface rocks. If you take the entire volume of the Earth, which is easy to measure, and the mass of the Earth, which we know from its orbit around the Sun, we get that the Earth has an average density—that is, mass divided by volume—of 5.25 grams per cubic centimeter. And if you pick up a bunch of rocks on the surface, and you put them on a scale, and you measure their volume, you find that surface rocks have an average density that ranges between 2.5 and a little more than 3 grams per cubic centimeter. If the whole Earth is about twice that dense, it's clear that the core of the Earth must be much denser than the rocks on the surface. This was the first clue that sedimentation—separation of the heavy elements from the light elements—had occurred.

Today, we know that the core of the Earth is largely iron, about 79% iron, 7% perhaps silicon, 5% nickel, and lots of other trace amounts of these heavy elements which settled out when the Earth was still in a molten state towards the core of the Earth. Thus, elements like platinum and iridium, gold, and silver are very rare on the surface of the Earth. And Alvarez thought that iridium might be a nice way to measure the sedimentation rates at the K-T boundary. Why? Because most of the iridium on the surface of the Earth comes from interplanetary space, riding in with the roughly 40,000 tons of space dust that settles onto the Earth each year. This dust has its origin in the origin of the solar system. Most of the material of the solar system in its early days was gathered up into the planets—Mercury, Venus, Earth, Mars, Jupiter, Saturn, Uranus, and Neptune—but there are little bits left over that have never managed to make it onto the surfaces of large planets. In particular, there's a large belt of rocks, ranging in size from a couple hundred kilometers down to the size you can hold in your hand, called the asteroid belt, between the orbits of Mars and Jupiter. There are millions of these rocks and they're constantly colliding with each other and producing dust. The Earth travels around the Sun each year through this cloud of dust and it gradually settles onto the Earth.

Now, 40,000 tons may sound like an enormous amount of space debris coming in every year, but if you calculate how much that means in a layer spread over the Earth, it's only about a couple billionths of a centimeter per year, and over the entire 4.5 billion years of Earth's history, if that sedimentation rate were constant, it would only add up to a layer about this thick. So we're not about to get inundated by this dust.

Nonetheless, it carries to Earth in the form of tiny particles, ranging in size from about $\frac{1}{100}$ of a millimeter, far too small to see, up to about a millimeter in size—grains of sand which actually burn up as they go through the atmosphere and produce what we call shooting stars, although their atoms still settle to the Earth. It carries to Earth a representative sample of the material of the solar system, undifferentiated, like on Earth. Thus, it contains the element iridium.

It's important to note that iridium-193 is not radioactive. The clock in Alvarez's scheme comes not from the fact that the atom is going to decay but from the fact that this infall rate of dust is very constant. When the ocean sedimentation rate becomes high, the concentration of iridium will be low. A lot of material gunk from the surface of the ocean settling to the floor will have come in over a short period of time and the amount of iridium that's come in in that time is small. Vice versa: When the sedimentation rate is very low, a small amount accumulating every year, or decade, or millennium, the amount of iridium accumulated in that time stays constant, and the iridium concentration will be higher.

What Alvarez found, in a layer of sediment in Gubbio, Italy, that was now raised above the sea, was that the iridium concentration jumped from 0.3 parts per billion to 7 parts per billion—by a factor of 25—precisely at the K-T boundary above which no dinosaur bones are found. There are 2 possibilities for this huge sudden change. Either there was a huge sudden change in the sedimentation rate or there was an extra influx of iridium from space.

He then went to a deposit in Denmark where the same level was exposed above the sea and another one in New Zealand and found that there, again, the iridium layer right at the K-T boundary jumped by a factor of as much as 130 from below the boundary to at the boundary. In 1980, this led him to publish a paper entitled "Extraterrestrial Cause for the Cretaceous-Tertiary Extinction." It was published in *Science*, and I suspect that the only reason it was published in *Science* was because in addition to his 2 coauthors, who had done the geological fieldwork and the laboratory analysis with him, he included his father, Luis Alvarez, who is a Nobel laureate in physics.

The paper proposed that a meteor impact had killed the dinosaurs. Catastrophism was back, and most scientists simply didn't believe it.

I remember distinctly a lecture in 428 Pupin, the room in which I lecture and the room in which Rabi introduced the atomic clock, where Louis Alvarez gave a talk on his paper in the spring of 1981. He made a fascinating, to me, case for the impact of a large object, 10 kilometers in size, which among other things, spread iridium dust over the entire Earth. I remember walking out of that lecture with people shaking their heads, saying, "Just don't believe it." Well, 30 years later, Alvarez was unquestionably right. The evidence for a meteor impact and its global consequences are overwhelming. In fact, late in 2008, just a couple of months ago, I attended another dinner at Columbia honoring Alvarez for his discovery, in which the citation went on at some length about how he had reintroduced the notion of catastrophes into legitimate modern science.

The iridium layer has now been found at the K-T boundary at over 120 different sites around the world, with a distribution of concentrations that steadily decreases as you go away from the Caribbean ocean, suggesting that that's where the meteor landed. Indeed, the crater made by the impact has now been found. As a consequence of oil exploration in Mexico, it was discovered that on the tip of the Yucatan Peninsula, there is a large region below the surface, now filled in by sediment, of fractured rock that extends down for several miles. This is now called the Chicxulub crater, after a word in the local Mayan language, and is, as Alvarez had predicted, roughly 200 kilometers in diameter, implying an asteroid of about 10 kilometers in size, hitting with an energy that's roughly 50 times that of the world's entire arsenal of 10,000-megaton nuclear warheads. It was a big bang.

The rock was vaporized to a depth of 2 kilometers. Nine hundred kilometers, 500 miles away, in Haiti, the ejecta, the material that rained down from this impact, is about half a meter thick. In addition to being rich in iridium, the ejecta layers also contain very unusual things: fused quartz; little hard spheres called tektites, things that are like glass; diamonds; and zircons. All of these things only form under extreme pressures. Diamonds are normally made miles down in the Earth with a huge overlying burden. In this case, they were made instantaneously as this huge rock hurtled towards Earth at 40 km/s and smashed into Mexico.

Zircons, it turns out, are particularly valuable for dating this event. In fact, you can use them to date events ranging from tens of millions of years ago to billions of years ago. Zircons are made up of the element zirconium, which is atom number 40 in the periodic table: 1 atom of zirconium, 1 atom of silicon, and 4 atoms of oxygen, and they form beautiful little crystals. They're valuable for dating because in the formation of the crystal, the element lead is specifically excluded, while the element uranium is readily substituted for some of the zirconium positions in the crystal lattice.

As a consequence, one can monitor the decay of uranium to lead. The method is particularly accurate because there are 2 different isotopes of uranium that are involved: uranium-238, the most abundant isotope, which makes up 99.3% of all uranium found on Earth today, and uranium-235, which makes up most of the remaining 0.7%. The difference in the abundance of these 2 isotopes is clear by their half-lives. Uranium-238 has a half-life of 4.5 billion years, just equal to the age of the Earth. So since the Earth's formation, just half of what was originally present has decayed away. Whereas uranium-235 has a half-life that's 7 times shorter, only about 700 million years, and thus, as one would expect, 7 half-lives have gone by, and so $\frac{1}{2}$ to the 7th power, or only about 1% as much, is left behind.

Uranium-238 decays to lead-206 through a complicated series of almost a dozen steps, and uranium-235 decays to the isotope lead-207. Thus, in a single crystal, one has 2 different, independent radioactive clocks. Given their precisely measured half-lives, we can predict at any point in time what the ratio of uranium to lead for each of these 2 isotopes should be, and we can plot that on a curve with the ratio of uranium-235 to lead-207 on one axis, and the ratio of uranium-238 to lead-206 on the other axis. We then can plot, by measuring in different points in the crystal the amount of lead and uranium remaining, what the ratio is.

If we assume there's no lead in the original zircon—a safe assumption given how it's formed—and that all the lead present comes from uranium, we could instantly get a date. But it's possible that some of the lead atoms produced in the decay of uranium have been leached out of the rock by various geological processes. Nonetheless, lead-207 and lead-206, being chemically identical, will have been leached out at just the same rate. And so by plotting the ratios of uranium to lead for each of the 2 isotope sequences, we get a perfectly straight line on the same diagram. Where that straight line intersects the theoretical curve of predicted half-lives will give us the age of the sample. These ages are accurate to better than 0.01%. Using a technique called SHRIMP, which stands for Sensitive High-Resolution Ion [Micro]Probe—scientists love cute acronyms—we can peel away the layers of the zircon crystal bit by bit and see how, over time, if it's been altered by geological processes. Nonetheless, this clever diagram of the theoretical curve versus the measured quantities of uranium and lead provides us an age precisely. The answer in the case of the zircons produced in the meteor impact that ended the reign of the dinosaurs is 64.5 million years, with an accuracy much better than 0.1%.

Direct evidence has also been found that allows us to reconstruct the immediate aftermath of the impact of this huge meteor with Mexico. Layers of sandstone created in a Caribbean-wide tsunami have been found from Alabama to Guatemala. It's a very special kind of sandstone, because unlike the normal kind, which is built up over hundreds of thousands or millions of years as layer upon layer of sand accumulates, there are no little channels through this sandstone which correspond to tunnels that little worms make as they wiggle through the floor of the ocean. This huge layer was laid down suddenly by a tsunami and solidified without the chance for biological disturbance.

In addition, we find a layer of soot all over the world from the massive forest fires that were ignited by the impact. The impact made such a shockwave, a sonic boom if you will, in the atmosphere that propagated over the entire Earth and was sufficiently potent to instantly ignite all of the lush foliage on Earth into enormous wildfires. The ashes from those fires are everywhere.

It turns out the dinosaurs were extremely unlucky as to where this meteor, which totally by chance, happened to hit in Mexico. The rock in Yucatan is rich in anhydrite, a gypsum-like rock. Gypsum is made of an atom of calcium, an atom of sulfur, and 4 atoms of oxygen: calcium sulfate. This rich sulfur mix, when vaporized by the impact and thrown into the atmosphere, mixed with water vapor, becomes H_2SO_4, sulfuric acid.

It turns out when sulfuric acid is thrown high into the atmosphere, as it is today sometimes by volcanoes—the Mount Pinatubo eruption in the early 1990s being the most recent example of this—it condenses into tiny little droplets which are extremely reflective of sunlight. It makes a reflecting blanket over the entire atmosphere of the Earth.

Sunlight, instead of coming in and warming the Earth, just bounces away, reflecting back into space, and temperatures on Earth apparently plunged from about 20° C, a sort of comfortable 70°, to −5° C, below freezing, in a matter of a week.

We even know the month in which this meteor hit because of this sudden freeze. A geologist working in the fossil beds of Wyoming discovered the fossil of a lily pad leaf from this period of time. Now, you know what the back of a leaf looks like: It looks like the branches of a tree, a central vein and smooth branches coming off it. These are the little conduits for water and substance in and out of the leaf. In this particular fossil, rather than the smooth flow of these lines, the geologist found a very jagged, fractured kind of look, and he was very confused as to what kind of plant would leave such a peculiar angular structure to its leaf-vein system, until he had an idea. He took a lily pad that was alive today and stuck it in his freezer overnight. When he took it out, he found the same angular breaks in the system of veins on the back of the leaf that he saw in this fossil. And, judging from pollen that was encapsulated at the time in the reproductive state of that leaf that he found in the fossil, he could tell that it was sometime in June, 64.5 million years ago, that the dinosaurs died.

In addition to the sulfuric acid thrown into the atmosphere, it doesn't stay up there forever, and it eventually falls out as acid rain. We're not talking about acid rain such as that produced by power plants today, which has caused the introduction of some of our pollution control laws; we're talking about lots of acid rain, pure sulfuric acid in the amount of about 10 kilograms for every square meter of soil. How, you might ask, could we possibly know the amount of acid produced in the atmosphere? We can estimate it from the size of the crater and the amount of sulfur that we think was there, but we have another record, a remarkable isotopic record left in the skeletons of the little forams in the top of the sea.

The ratio of strontium-87 to strontium-86, as I've told you before, is a critical one for dating and tracing isotopic signatures, and strontium and calcium are homologues, and so strontium is easily incorporated into the shells of these creatures. The strontium-87–to–strontium-86 ratio in the ocean—the ocean being a very big place—is pretty stable, and it goes along at a steady level for a long period of time, right up until the K-T boundary, at which point there's a sudden jump in the strontium-87–to–strontium-86 ratio. What could cause this? Acid rain.

It turns out the strontium-87–to–strontium-86 ratio of rocks on land, the continental rocks, is slightly higher than that of the strontium-87–to–strontium-86 that's dissolved in seawater. But this deluge of sulfuric acid pouring out of the sky after the meteor event actually eroded the rocks on land sufficiently, which running into the rivers and out into the sea, changed the strontium-86–to–strontium-87 ratio of the entire ocean, sufficient to have it reflected in the foram shells that we see in ocean cores today. Quite remarkable.

Longer-term impacts came from alteration of the atmospheric composition. After the 100 trillion tons or so of dust and acid rain fell out of the atmosphere, probably over a few years, sunlight could reach the surface again. The huge amount of water vapor that was thrown up by the explosion—it landed in part in the Caribbean Sea, after all—in addition to the huge amount of carbon dioxide released when the limestone in the area was vaporized and the huge carbon dioxide release from the forest fires which raged all over the Earth, caused an enhanced amount of CO_2 in the atmosphere. And extra CO_2 in the atmosphere leads to extra heat-blocking powers of the atmosphere, meaning the temperature rose. The greenhouse effect raised temperatures for about 10,000 years. We'll talk more about the greenhouse effect in a few lectures.

There's excellent fossil evidence of massive changes in life at the K-T boundary. There was a massive die-off of plants; about 57% of all species seem to have gone extinct. Immediately thereafter, there was a blossoming of fungi. Why? Because fungi don't need photosynthesis, so they can thrive in the dark, as you know, feeding off the decaying matter from all the plants that died when the Sun went away. This was followed by a huge increase of ferns, an effect that we've seen recently in the eruption of Mount St. Helens. The eruption of Mount St. Helens, the volcano in Oregon and Washington, flattened all of the material and killed all plant life on the sides of the mountain. The first species to be reintroduced are ferns, and we see this in the fossil record from the spores these ferns left behind right above the K-T boundary.

The species of amphibians and birds largely appear to have survived this, but the big winner was the mammals. Mammals came into existence before the K-T boundary. They started diversifying about 30 million years before this catastrophic event. By the time of the K-T boundary, however, most mammals were really pretty small, sort of rat-sized creatures. But because of their small furry nature and, perhaps, their ability to store up nuts for the winter, they survived this terrible asteroid-induced winter; unlike the massive dinosaurs, who needed to eat huge amounts of plant life every day—when all their plants died, they went with them. As the dinosaurs disappeared, ecological niches opened up for the mammals to thrive. And though their extreme diversification, which we see on Earth today, didn't happen for several million years, the opportunity was there, and mammals now dominated the Earth.

About 6 million years ago, an early line emerged from the hominid family, from the great apes: chimps and bonobos, our closest relatives. Using molecular clocks, we can date this event, consistent with the fossil record in which the ages are determined from radioactive dating of the associated rocks. Many species emerged and evolved from this family—*Homo erectus* about a million years ago, and *Homo neanderthalensis* about 400,000 years ago. But both of these species lost out after the emergence of *Homo sapiens*, us, about 150,000 years ago. And here we are, 150,000 years later, only 0.003% of Earth's life, piecing that history together again, atom by atom.

The Cretaceous-Tertiary meteor, the K-T meteor, clearly had a global impact on climate for many millennia after the event, and it profoundly changed the balance of life on Earth. It was, however, neither the first such impact, nor was it the only time that a mass extinction had wiped out a large fraction of all species on Earth. Indeed, an event of unknown origin killed more than 90% of all species on land and in the ocean about 251 million years ago. Whether this was caused by an asteroid or not, we don't know. The surface of our Earth is constantly changing because of the erosion of wind and water and because of the wandering of the continents, earthquakes and volcanoes, the effects of plate tectonics. So that meteor crater may be lost forever.

There were at least 3 other mass extinctions that have occurred over the last 500 million years since life diversified in the Cambrian explosion, and actually, a fourth is now underway. The extinction rate today of all species—of plants, animals, and ocean-dwelling creatures—is estimated to be at least 100 times the average rate over the last 500 million years. And, of course, we know who's responsible for this latest mass extinction. The history of life has been a bumpy ride. Next time, we'll trace that history, from the interstellar clouds where the molecules of life collected to us.

Lecture Fourteen
The Origin and Early History of Life

Scope:

One of the more remarkable things about our planet is how rapidly life formed—namely, in the first 10% or less of the solar system's lifetime. In this lecture, we begin by examining molecules discovered in interstellar clouds of the type from which Earth formed and find that all the basic building blocks for life are present. We then explore the remarkable fact that all amino acids—on Earth *and* in meteorites—are "left-handed" and speculate as to how this came to be. Geological evidence then allows us to trace the development of life from single-celled organisms, which forever changed our atmosphere's composition, to the Cambrian explosion just more than 500 million years ago, during which life blossomed in all its complex forms to cover the Earth.

Outline

I. The precursor molecules of life are abundant in interstellar space.
   A. As we shall see in more detail in upcoming lectures, new stars (and the planets that frequently accompany them) are born in clouds of gas and dust in interstellar space.
   1. In most of the volume of interstellar space, there is only 1 atom for every cubic area 10 cm on a side, and the atoms are moving very quickly, characterized by a temperature of 1 million degrees.
   2. Cold regions form when this gas eventually cools, and much denser regions arise, where the number of atoms reaches thousands per cubic centimeter.
   3. The excess mass in these denser regions attracts more mass, so they get denser and colder still, eventually reaching the point where the mass of thousands to millions of stars is concentrated in a cloud only 10 degrees above absolute zero.
   B. So many of the atoms in these clouds are bound together in molecules, with H_2 predominating, that we call these structures molecular clouds.
   1. These clouds are filled with tiny dust particles, on the surfaces of which atoms come together to form molecules.

2. We can detect and count these molecules through the microwave radiation they give off as they make transitions between rotationally and vibrationally excited states.
3. More than 140 different molecules have been discovered to date, most of which contain the common elements hydrogen, carbon, nitrogen, and oxygen, but also including silicon (Si) and sulfur (S).
4. CO is the most common molecule after H_2, but water (H_2O) and ammonia (NH_3) are also prevalent.
5. Many of the molecules are carbon-containing organic molecules present in all life on Earth, such as ethyl alcohol (C_2H_5OH), formaldehyde (CH_2O), and acetic acid ($C_2H_4O_2$).
6. The most complex molecule discovered to date has 13 atoms, but many radio-frequency lines signaling the presence of other molecules remain unidentified.
7. It is clear the raw materials for life are present in all newly formed solar systems.

II. Meteorites show that even more complex organic molecules came together in the early solar nebula as the planets and Sun were forming.
   A. Analysis of meteorites that have fallen to Earth show that they contain the purines (adenine, guanine) and pyrimidines (cytosine, thymine) that form the basis of DNA's code.
   B. They also contain a number of the amino acids that are the building blocks of proteins, the basic molecules of life.
   C. A remarkable fact is that these molecules—in meteorites and in all Earth-based life—are left-handed.
      1. Many complex organic molecules come in 2 varieties that are chemically identical (they contain exactly the same number of each kind of atom).
      2. The sole difference is the 3-dimensional arrangement of the atoms: One is the mirror image of the other.
      3. Since one variety bends light one way and the other bends it in the opposite direction, they are dubbed left- and right-handed molecules.
      4. The amino acids found in all forms of life on Earth are 100% left-handed; there is also a marked preference for left-handed molecules among those found in meteorites, the most ancient objects in the solar system.

- **D.** It is possible to speculate on the origin of this asymmetry.
    1. As we shall see, there is evidence that a massive star exploded in our vicinity just as the solar system was being born.
    2. Such explosions often leave behind neutron stars, among the most bizarre objects we know of in the universe.
    3. Such stars produce circular waves of light that could sweep over the young solar system and selectively destroy molecules formed in a right-handed way.
    4. My Columbia chemistry colleagues have shown how a small initial asymmetry could be amplified to left-handed dominance under the conditions expected to prevail on the early Earth.

III. The young Earth had conditions conducive to the formation of life.
- **A.** We begin with a very simple definition of life: a self-reproducing chemical system capable of Darwinian evolution.
    1. It may not sound like your favorite cousin, but it is a good bare-bones description.
    2. Life clearly must reproduce or we wouldn't consider it life.
    3. Science fiction writers have imagined non-chemical-based systems to be alive, but ours is a conservative hypothesis.
    4. Evolution involves a means of transmitting instructions from one generation to the next, which can include errors in transmission—and therefore can be subject to natural selection.
- **B.** This definition requires that molecules store information, and therefore they must be reasonably complex.
    1. Complex molecules require a moderate temperature—too cold and the atoms never meet up to join together; too hot and they are instantly broken apart by collisions.
    2. Complex molecules are most easily formed in a liquid state, in which atoms meet frequently but are not rigidly locked in place.
    3. Liquid water is made of the 2 most abundant chemically active atoms in the universe and has particularly copacetic characteristics but is *not* essential for forming complex molecules.

4. Abundant liquid water on Earth clearly aided the development of life.

IV. Life's history is one of adaptation to changing conditions, some of which were brought about by the existence of life itself.
   A. The Earth was initially far too hot for molecules to survive.
   B. As it cooled and the crust formed, molecules rained down from comets and asteroids, striking the surface as the young solar system cleaned up the omnipresent disk of debris.
   C. The Sun was less bright than today, but the atmosphere, rich in CO_2 from numerous volcanoes, kept the temperature above freezing, so liquid water rained out of the nascent atmosphere.
   D. Within less than 750 million years, life emerged.
      1. The oldest actual fossils are 3.5 billion years old (from 1.0 billion years after the Earth formed), but stromatolites, the layered rocks formed from the remains of blue-green algae (cyanobacteria), date from almost 4 billion years ago.
      2. As we learned in discussing the history of agriculture, living things prefer carbon-12 to carbon-13, so rocks with a high carbon-12/carbon-13 ratio provide evidence for earlier life.
      3. Such rocks with life-like carbon ratios have been found on Akilia Island, Greenland.
      4. The rocks have been dated using uranium-lead dating in zircons. Zircons originally contain no lead, and since uranium decays to lead, the ratio of uranium to lead dates the rocks as an accumulation clock.
      5. There is some dispute about the exact age of these rocks, the oldest found on Earth, but it is apparently between 3.7 and 3.85 billion years—implying life was already prevalent a mere 700 million years after Earth formed.

V. Life took about 3.8 billion years to get to us.
   A. It should be noted that our Sun has still not lived half its life.
   B. We contain roughly 50 trillion cells, suggesting that, averaged over the time since life began, we have been adding 13,000 cells per year.
      1. The rate has been far from uniform, however.

      2. The first multicellular organism took at least 2.5 billion years to evolve.
- **C.** Subsequent evolution produced an elaborate feedback loop between the changes in life and the changes that life effected on the Earth's atmosphere.

Suggested Reading:

Dawkins, *Climbing Mount Improbable*.

Lane, *Power, Sex, and Suicide*.

Thomas, *The Lives of a Cell*.

Questions to Consider:
1. Does the presence of appropriate raw materials and a relatively benign environment make the emergence of life inevitable?
2. How might other definitions of life, or embellishments to my definition, lead to radically different creatures on other planets?

Lecture Fourteen—Transcript
The Origin and Early History of Life

By the time the Chicxulub meteor put an end to the age of the dinosaurs, life had already been flourishing on Earth for more than 3.5 billion years. Indeed, in the first 10% of the life of our parent star, the Sun, life emerged in abundance on this planet.

Why did this happen so fast, and why has it not emerged elsewhere? The answer to the first question involves the atomic raw materials that are available and the molecule-level conditions which obtained on the early Earth. The answer to the second question, why hasn't life emerged elsewhere, is: maybe it has.

The precursor molecules of life are abundant in interstellar space. As we shall see in more detail in upcoming lectures, new stars and the planets that frequently accompany them are born in clouds of gas and dust in interstellar space. In most of the volume of interstellar space, there is only about 1 atom—and remember how small an atom is—for every cube about 10 centimeters on a side, about the size of this Kleenex box. There's 1 atom in there, and 1 atom over there, and there's 1 atom over there. That's really empty space, because in this box right now, just full of air, there are about 10^{21} atoms in the air in this room. So interstellar space is really empty.

Furthermore, in most of that volume, the temperatures are extremely high, about a million degrees, heated by the explosions of stars over the course of the life of the galaxy. So the individual atoms that are there, mostly hydrogen, of course, are running around with extreme rapidity. When they collide with each other, they knock their electrons out of each other's orbit, and the material is, by and large, that fourth state of matter, which we don't see often on Earth, called a plasma.

There are regions of space that manage to cool off, and where the gas gets cooler, it starts to collect. Denser regions arise until the atoms reach densities of thousands of atoms per cubic centimeter. A sugar cube is about the size of a cubic centimeter, and rather than 1 atom in a space this large throughout most of the hot regions of space, there are clouds where there are a thousand atoms inside this sugar cube. Now, again, this sugar cube contains more than a trillion trillion atoms at the moment, but relative to earthbound things, space is still empty—but this is a dense region of space.

As the matter gets denser, that means there's more mass. When there's more mass, there's more gravitational attraction, and so the little clump of matter over here that's managed to cool itself down, have its electrons join up with its atoms, become a gas rather than a plasma—the atoms exerted gravitational attraction from material over here, which falls slowly towards it, and material over here, which falls slowly towards it. That, of course, just creates a larger mass here, which has an even greater gravitational force, which extends farther out into space so that mass here, and here, and here all starts falling towards this region, as well.

Gradually, over the timescale of millions of years, these cold regions can accumulate the mass of thousands to millions of stars, huge amounts of matter, concentrated in a cloud that's only about 10° above absolute zero. In other words, the atoms are moving around very, very slowly. So many atoms are there and they're moving so slowly, that they tend to bind together to make molecules. We call these structures in space molecular clouds. And, of course, since hydrogen is the dominant element in the universe and forms molecules easily by linking 2 of them together, the vast majority of these clouds are made of 2 hydrogen atoms stuck together, H_2. These clouds are also filled with tiny particles of dust, far too small to see with the naked eye, which range from collections of tens of thousands of atoms that are all stuck together in a quasi-solid to really little grains that you could almost see made of carbon, and iron, and other such elements.

On the surfaces of these dust grains is a place for atoms to meet up. The dust grains—because they get bombarded occasionally by the same cosmic rays that produce radioactive isotopes in the atmosphere of the Earth—spew off electrons, and so the grains become positively charged. Atoms that stick to them, then, move around on the surface and, eventually, can find each other to link up. In interstellar space, even though the cloud is dense, there's still huge space between the atoms. In the air in this room, the atoms are about this far apart if each fist represents 1 atom. In these molecular clouds, the 2 fists would be blocks apart, and so the odds of them coming together and coming together slowly enough that they just gently merge and share their electrons to make a molecular bond is small. On the surface of these dust grains, however, as they vibrate around on the surface, they can meet up with each other and link together because, first of all, they're moving more slowly, and secondly, they have a solid surface on which to rest.

We detect these molecules formed on these dust grains and subsequently ejected back into the gas through the microwave radiation they emit. In addition to the jumping between electron levels of an atom—which emits mostly visible light that our eyes can see—these molecules, with 1, 2, 3, 4, 5, 6, up to 15 atoms together, manage to undergo 2 other kinds of excitations.

One is they vibrate. You can imagine them as 2 little balls connected with a spring, and given a little nudge from a passing atom, they'll vibrate like this. Now, just as the electrons orbiting the nucleus of the atom can only exist in specific orbits and, therefore, only have specific energies, the transitions between which give rise to light, these little vibrating springs can only vibrate at certain frequencies—this fast, or twice as fast, or 4 times as fast—and so also give off only certain frequencies of light, giving a unique signature to that molecule.

Finally, there's a third mode of even lower-energy excitation, in which if a molecule gets bumped into, it can sort of tumble, rotate around. And, again, it can't rotate at *any* speed; it can rotate this fast, or it can rotate twice that fast, or it can rotate 4 times that fast, but not at speeds in between, again giving a unique signature to each molecule.

These types of excitations, vibrational excitations and rotational excitations of molecules, give off radiation between the infrared octaves of the spectrum into the radio octaves of the spectrum. In other words, long-wavelength light, because they are much lower energy than the little electrons which are zipping around the nucleus of the atom. When the electrons change levels, they give off high-frequency light, light you can see. When these vibrational modes or rotational modes give off energy, it's much longer wavelength light that only radio telescopes or infrared telescopes can pick up.

Nonetheless, the important point is that each different kind of molecule—water, carbon dioxide, or whatever—has a unique set of modes in which it can vibrate and rotate, and therefore, it gives off a unique set of specific frequencies which, when measured with a telescope, can identify that molecule. Over 140 different molecules have already been discovered, most of which contain the most common elements, unsurprisingly—hydrogen, carbon, nitrogen, and oxygen—but also including things like silicon and sulfur. After H_2, the most common molecule is carbon monoxide, carbon and an oxygen atom joined together, and its radiation dominates these interstellar molecular clouds. But water and ammonia, 2 common elements on Earth, are also present.

Many of the molecules are carbon-containing organic molecules present in all life on Earth, such as ethyl alcohol, formaldehyde, and acetic acid. These exist in enormous quantities in these molecular clouds in interstellar space. The most complex molecule discovered to date has 13 atoms joined together. But there are many signals in the radio part of the spectrum which signal the presence of molecules but remain unidentified because we haven't synthesized those molecules in the lab and made them vibrate and rotate to reveal their secret wavelengths. It's clear, however, that the raw materials for life are present in all newly formed solar systems, because all such solar systems emerge from these molecular rhythm clouds.

The meteorites in our solar system, the rocks left over from its formation which fall to Earth, show that even more complex organic molecules came together in the early solar nebula as the planets and the Sun were forming. Analysis of the interiors of meteorites which have fallen to Earth shows that they contain purines and pyrimidines. Purines are 5 carbon atoms, 4 hydrogen atoms, and 4 nitrogen atoms joined together and are 2 of the 4 chemical bases of DNA, adenine and guanine, the genetic code of life. And pyrimidines are the other code; they are 4 carbons, 4 hydrogens, and 2 nitrogens, and they form cytosine and thymine, the other 2 letter codes of DNA, plus uracil, the code in RNA for translating the DNA instructions out to the rest of the cell. Meteorites also contain a number of the 20 amino acids which are the basic building blocks for all proteins and, therefore, the basic molecules of life.

A remarkable fact is that these molecules in meteorites and in Earth-based life are left-handed. What do I mean by "left-handed"? Many complex organic molecules come in 2 varieties which are chemically identical; that is, they contain exactly the same number of each kind of atom. The sole difference between them is the 3-dimensional arrangements of these atoms. One is the mirror image of the other. There's nothing you can do to fit a right-handed glove onto your left hand. You can turn it around, you can turn it upside down, but there's no way you can make it match perfectly. Likewise, when you hold up your hand to the mirror, you get the reverse image. A mirror-image molecule is one that's the same: You hold it up to a mirror, you get the reverse image. There's no way you can rotate the molecule so it looks the same. In fact, the same atoms are there, but they form a mirror image of each other.

It turns out that these molecules are called right- and left-handed because of the way they treat light. Light, as we learned in one of the early lectures, is a wave. It's a wave of electric and magnetic fields, and if I were to try to reproduce that with my hand it would look something like this. The electric field grows and falls, and in response, the magnetic field grows and falls in a perpendicular direction. So a propagating light wave looks something like this. My students have made a little dance out of that routine. If we follow just the electric component of the field, the wave goes up and down like the wave of water, but there's no reason that a molecule emitting such light needs to go up and down in this plane. It can go up and down in this plane, it can go up and down in this plane; there is an infinite number of rotations of that plane of oscillation for the electric wave.

Polarized light means light which has restricted the modes in which the wave oscillates. Light that is purely polarized means that all the waves are going up and down in this direction, or alternatively, all the waves are going back and forth in this direction in terms of their electric field motion. Polarized sunglasses work by providing a little picket fence in the plastic of the glasses such that the light, say, reflecting off the hood of your car, that's polarized in this direction comes in and sort of buries itself in the hood of the car and so doesn't shine off and glare in your eyes. The light that's coming in polarized in this direction *does* bounce off the hood of the car very easily and would come to your eyes and make a blinding glare unless you had a little picket fence in which the light could clatter against and not get through. Polarized sunglasses, then, pick out a particular orientation of the oscillation of the waves of light.

Right- and left-handed molecules also pick out particular planes of polarization in just the same way, and if you have a little vial of right-handed molecules, in which the polarized light is going this way, it will actually rotate that plane of oscillation to the right as it passes through, whereas left-handed molecules cause it to rotate to the left. The name, right-handed and left-handed molecules, is because they rotate polarized light in opposite directions.

All of the amino acids, all 20 of them, which have a handedness—some of them don't have a handedness, but many of them do—that are found in all forms of life on Earth, are left-handed—all of them, left-handed. You would think, naively, if you just shake up some molecules in a test tube before life exists, you'd get an equal number of right-handed and left-handed ones. But life on Earth today consists of only left-handed amino acids.

Remarkably, when we look in the meteorites for the amino acids that are there, we also find a preponderance of left-handed molecules. These are the most ancient objects in the solar system, and somehow, they have already had imprinted on them the notion that molecules are preferentially left-handed. It's possible to speculate on the origin of this asymmetry. I find this a fascinating and delightful speculation, but I can't argue that it's unquestionably true. As we shall see, there is good atomic evidence that a massive star exploded in our vicinity just as the solar system was being born. The radioactive isotope aluminum-26—which, when it decays, on a relatively short timescale, turns into magnesium-26, an isotope we find a lot of in the solar system—is prevalent. The only way you can produce this short-lived radioactive isotope is in the explosion of a massive star.

Such explosions often leave behind bizarre objects called neutron stars, where the mass of an entire sun has been compressed into something the size of Manhattan, 10 kilometers across, the size of the meteor that wiped out the dinosaurs. That means the atoms—and, actually, the atoms are crushed out of existence, so the protons and neutrons—are packed together to such a density that the entire object has the density of the nucleus of an atom, 1 billion tons per teaspoonful of matter. This sugar cube would weigh a billion tons.

These bizarre objects have a number of properties which we'll explore in a later lecture, but one of them is that they spin very rapidly, and, because of intense magnetic fields, send out, along their north and south magnetic poles, spirals of circular waves of light, on the north pole rotating in one direction and on the south pole rotating in the opposite direction. When molecules that have a handedness absorb light that's going in the same direction as they are, they're excited. When they accept light going in the opposite direction, they tend to be destroyed. Right-handed molecules are destroyed by left-handed circular polarization, and that can produce an excess of lefties, say, 10%. My Columbia chemistry colleagues have actually shown how a small initial asymmetry could be amplified to left-handed dominance under conditions that are expected to prevail on the early Earth. In this scenario, a star explodes and leaves a neutron star behind. The light from one pole of that star irradiates the cloud that will become the solar system and produces a preference for left-handed molecules, perhaps an excess of 5 or 10%.

My colleague Ron Breslow and his student Mindy Levine have exposed 5% excess left-handed amino acids to the hot, dryish conditions, with a little bit of water, that are probably appropriate to the early Earth. And they have found that the left-handed molecules and the right-handed molecules in the solution tend to bind together, and crystallize, and fall out of the solution, leaving only the excess, the 5% excess, of left-handed molecules free in the solution to wander around and connect to other molecules, making them more complex, things perhaps like proteins.

This is a fascinating notion, that the explosion of a star and the irradiation of our nascent cloud with circularly polarized radiation could have produced a small preference for left-handed molecules, and then the conditions of the early Earth could have amplified that preference to the point that every amino acid on Earth today is left-handed. The fact that the left-handed preference is also seen in the meteorites gives some credence to the story.

The young Earth had conditions that were reasonably conducive to the formation of life. What do I mean by "life"? We have to begin with a very simple definition of life, stripped to its bare bones. I define life as a self-reproducing chemical system capable of Darwinian evolution. That may not sound much like your mother-in-law—or it may sound like your mother-in-law—but it's a good bare-bones description of what life is like.

Clearly, we would all agree that life must be self-reproducing. That's an essential element of life, that it reproduces itself. Rocks don't do that; bacteria and humans do. Secondly, my suggestion is that life is a chemical system. Now, science fiction writers have imagined non-chemical–based systems to be alive. For example, the famous astrophysicist Fred Hoyle, who doubled as a science fiction writer, wrote a wonderful book called *The Black Cloud* in the 1950s, which as in most good science fiction, was more about the humans than the creatures from afar. But the thing that was alive in that story was a large dark cloud, perhaps one of these interstellar clouds diffused with dust, whose intelligence was conferred by the threaded magnetic field through the cloud. I'm not saying it's inconceivable that there could be a kind of life, even intelligent life, that's not based on chemical molecules, but excluding that from our definition is a conservative hypothesis.

It's important to remember that chemistry is identical everywhere in the universe. The atoms of carbon that exist in my thumbnail are identical to the atoms of carbon in the most distant quasar I can see across the universe. Furthermore, the rules under which those atoms of carbon join together to make chains or join with oxygen, or nitrogen, or other kinds of atoms to make complex molecules are identical across the universe. So when I say life is based on chemistry, I don't have some chauvinistic notion of our chemistry or my chemistry; it's *the* chemistry of the universe, and it will operate on the early Earth the same way it operates in the most distant galaxy we can see.

So, self-reproducing and chemical—the third component of life is that it's capable of undergoing evolution, means it must transmit instructions from one generation to the next in a mechanism that allows mistakes, errors in transmission, which can therefore be subject to natural selection and, therefore, change, speciation. This definition requires that molecules store information. If they have to pass it on to the next generation, they surely have to store information in some way. Therefore, the molecules must be reasonably complex; H_2, 2 hydrogen atoms joined together, or even H_2O for that matter, an oxygen joined to 2 hydrogens, is not very complex. It can't store a very large amount of information, and putting a lot of them next to each other doesn't help very much. The actual molecule that stores information in living creatures on Earth, of course, is DNA, which is a rather complex molecule consisting of thousands of atoms joined in very specific patterns which can then transmit information to the next generation.

This assumption of complex molecules requires that we need a moderate temperature. Temperature, you will recall, is just a measure of the speed with which atoms or molecules move. If it's too cold, the atoms move so slowly that they rarely meet up with each other and join together. If it's too hot, on the other hand, as soon as a molecule forms, it'll be smacked by another molecule, which then instantly breaks it apart, and so complex molecules will not evolve.

Complex molecules are also most easily formed in the liquid state. Recall in a solid that atoms and molecules are locked in place; they can't move. And if they can't move, they can't migrate to join up with other friends and build something complex. In a gas, the molecules are free to run around, but they encounter each other rarely, and when they do, they tend to bounce off each other like billiard balls. Liquid is a reasonable compromise between these 2 states. In a liquid, the atoms are touching, or the molecules are touching, so they're close to each other—they find lots of friends—but they're free to slide over each other and, therefore, to migrate through the liquid.

Water, liquid water, is made of the 2 most abundant chemically active atoms. The most abundant element in the universe, hydrogen—the second is helium, but that can't make molecules with anything, so cross that out—and the third most abundant atom is oxygen. So the 2 most abundant chemically active ones join together, with 1 oxygen and 2 hydrogen attached.

It's not true that water is absolutely essential for life. Any liquid would allow this conjoining of molecules that can migrate through it, but water is particularly copacetic because of many of its characteristics. In particular, water is said to be a highly polar molecule. That is, the distribution of electrons orbiting around in the oxygen and 2 hydrogen atoms—looks a little bit like a Mickey Mouse hat, with the big oxygen here and the 2 little hydrogens acting as ears—is not smoothly distributed. The positively charged nuclei are locked in place, but the electrons spend a lot more time orbiting around the oxygen than they do the hydrogen. Just this molecule on one side tends to be negatively charged, on the oxygen side, and where the 2 ears are, since the electrons don't spend much time with the hydrogen atom, it tends to be positively charged. Negatively charged down here, positively charged at the top.

This means that molecules that are wandering through, which have a slight charge of their own, or perhaps even ions that have lost an electron and, therefore, are positively charged, will preferentially stick to the negatively charged part of the water molecule. When you put lots of water molecules together, you can then imagine them as little scaffolds on which more complex molecules can be built because of this essential polarity.

Abundant liquid water on the Earth clearly aided the development of life. Life's history is one of adaptation to changing conditions, some of which were brought about by the existence of life itself. The Earth was initially far too hot for molecules to survive. As it gradually cooled over the first 100 or 200 million years or so, the crust formed on the surface, and molecules rained down from many asteroids, not once every 100 million years or so, like we have today, but perhaps every day, raining down onto the surface. In addition, comets, which are sort of dirty snowballs, balls of ammonia and water ice which contain within them the products out of which the solar system formed, including all these molecules we see in the net molecular clouds out of which stars form, are falling down out of this debris disk that surrounds the Sun; gradually, the Earth is sweeping it up, but all of the stuff is falling onto the surface.

The Sun was actually less bright than it is today, less luminous. It was putting out less energy. But the atmosphere of the Earth was very rich in carbon dioxide, even more than the amount of carbon dioxide we humans have put in it today, from the numerous volcanoes that were constantly erupting on this still-hot Earth. And that, as we'll see in the next lecture, provides a blanket which keeps the temperature warmer than it would otherwise be. And so the temperature was kept above the freezing point of water so that liquid water rained out of the original atmosphere.

Within about 750 million years, life emerged. The oldest actual fossils we have of life are about 3.5 billion years old. That is only a billion years after the Earth formed. The first solid evidence we have for life is blue-green algae, blue-green algae called cyanobacteria. They were the first to use photosynthesis and produce oxygen. The oxygen was poisonous to them, as a matter of fact, but it allowed the development of all the life that breathes oxygen that we have on Earth today.

These blue-green algae collected into vast mats living in symbiosis with bacteria and other primitive life that was around at the time. And we see their evidence today in geological formations called stromatolites, which are layered—intricately layered and folded—regions of minerals that were collected by these mats of algae and layered in place as they died. All of these organisms, the bacteria and the cyanobacteria, were cells that had no nuclei. They lived over most of the 4 billion years of Earth's history but now are found in only a few isolated places. They're also responsible for the geological deposits called banded-iron formations.

Iron, emerging from the interior of the Earth, where we learned most of it is, is initially soluble in water, but when it joins up with oxygen, it rusts. In other words, it precipitates out from the ocean. Banded deposits result from the periodic rise and fall of the oxygen levels in the water. Whether this was seasonal or some longer-term event, we don't know because we don't have a clock to measure 3.8 billion years ago. But these banded-iron formations were laid down between the period 3.8 billion years ago and 1.8 billion years ago, and they form the principal iron-ore deposits that we mine today.

We've learned in discussing the history of agriculture that living things prefer carbon-12 over carbon-13. So rocks with a high carbon-12–to–carbon-13 ratio provide evidence for early life. Such rocks with lifelike carbon ratios have been found on Akilia Island off Greenland, for example. The rocks have been dated using the uranium/lead dating in zircons that we talked about last time. The zircons, originally containing no lead, provide an excellent clock with the double uranium-235 and uranium-238 decay chains to give us a check.

There is some dispute about the exact date of the age of these Greenland rocks, the oldest ones found on Earth, but apparently, it lies somewhere between 3.7 and 3.85 billion years—close enough, as far as I'm concerned—implying that life was already prevalent only 700 million years after Earth formed. It then took another 3.8 billion years to get to us.

It should be noted that we're here, but the Sun has not yet lived even half its lifetime. We contain 50 trillion cells. If you average that out over time, it means we've been adding about 13,000 cells per year, but the rate has been far from uniform. For the first 2.5 billion years, there were not even cells with a nucleus. Multicellular organisms took another half a billion years to evolve. Subsequent evolution produced an elaborate feedback loop between the changes in life and the changes that life effected on Earth's atmosphere, as we'll discuss next time.

Life emerged on Earth from commonplace interstellar chemicals after less than 10% of our Sun's lifetime was complete. For billions of years, it was dominated by the physical conditions in which it found itself. Then, life gradually took over the planet. In the next lecture, we'll see how life also took over the atmosphere.

Lecture Fifteen
The History of Earth's Atmosphere

Scope:

The Earth's atmosphere is not in equilibrium. The large amount of free oxygen present (~20%) is a consequence of life on the surface. The Earth was formed in an environment where 98% of the atoms were hydrogen and helium; the Sun and Jupiter retain that composition today, yet both elements are nearly absent in Earth's atmosphere. Clues from our sister planets and from the geological record allow us to reconstruct the long-term history of Earth's atmosphere and the dramatic differences in climate—from tropical forests at the poles to a snowball Earth almost completely encased in ice—that these different atomic compositions produced.

Outline

I. The composition of a planet's atmosphere is determined by 3 factors: the strength of gravity at its surface, the surface temperature, and the raw materials available to make molecules.

  A. We should begin by asking the simple question: Why doesn't the atmosphere fall down?
    1. It is made of atoms, and atoms have mass.
    2. The planet's gravity attracts everything with mass toward its center.
    3. So why aren't all the air atoms lying on the ground?

  B. Air molecules *do* fall down—they are attracted to the planet, as are all other objects possessing mass.
    1. However, the thermal energy the atoms have—running around and bumping into each other because they are warm—keeps the atmosphere up.
    2. Not all atoms have the same speed, but for a given temperature, both an average speed and a distribution of speeds can be calculated.
    3. The average speed of an air molecule at room temperature is about 450 m/s; at that speed, a typical nitrogen molecule traveling straight up would reach 10,000 meters (the height of Mount Everest) before falling back to Earth.

- **C.** Since the temperature determines the average speed and the gravity determines the attractive force, these 2 quantities are critical for determining the thickness of the atmosphere.
- **D.** These 2 quantities are also critical for determining the atmosphere's composition.
    1. "What goes up must come down" is a falsehood (or else how could we send probes to the other planets and out of the solar system?).
    2. Every body has an "escape velocity" that, if you exceed it, will let it leave the planet forever and not "fall down."
    3. For Earth, the escape velocity, determined only by its mass and radius, is just more than 11 km/s.
    4. Thus any molecule in the atmosphere that exceeds this speed can escape.
    5. The speed of a molecule, as we learned when we discussed C_3 and C_4 plants, is dependent on its mass; the light ones move quickly, and the heavy ones move slowly.
    6. Since nitrogen molecules have a mean speed of 0.45 km/s (more than 20 times lower than the escape velocity), virtually none of them escape.
    7. However, hydrogen molecules have a mean speed of 1.7 km/s, so those moving only 6 or 7 times the average speed will escape.
    8. Thus, although the solar system formed out of material that was 98% hydrogen and helium, virtually none of those are left in Earth's atmosphere today.
    9. Mercury, which is small (low gravity) and close to the Sun (hot), has no atmosphere at all, as everything escapes.
    10. Jupiter, which is huge (high gravity) and far from the Sun (cold), is composed mostly of hydrogen and helium.

II. The composition of the Earth's atmosphere today is a consequence of these fundamental forces, which shape all atmospheres and Earth's particular history.

- **A.** Our atmosphere today is mostly nitrogen and oxygen.
    1. N_2 and O_2 molecules—not individual atoms—make up more than 98.7% of the atmosphere.
    2. Another 0.93% of the atmosphere is the noble gases (helium, neon, argon, and krypton), which never participate in chemical reactions.

3. The third most abundant molecule is water, which contributes 0.25% on average.
4. Carbon dioxide, methane, and nitrous oxides make up the remaining 0.04%, but despite their small fractions, they, along with water vapor, control Earth's temperature through the greenhouse effect mentioned earlier.

B. It is unsurprising that most of the molecules contain the elements hydrogen, carbon, nitrogen, and oxygen, since these are 4 of the 5 most abundant elements in the universe and, along with helium, constituted 99.6% of the mass of the cloud out of which Earth formed.

C. The other planets all have atmospheres that are dominated by these same atoms, although Earth is unique in containing a substantial fraction of free oxygen.
1. Oxygen is highly reactive (e.g., burning and rusting are a consequence of other atoms and molecules combining vigorously with oxygen).
2. Its continued (and unique) presence in our atmosphere is a consequence of life.

III. The initial composition of the Earth's atmosphere was very different from that in evidence today.
A. The initial atmosphere was almost certainly composed of the primordial elements, hydrogen and helium.
1. In addition, the compounds containing the most abundant other elements combined with hydrogen—ammonia (NH_3) and methane (CH_4)—were likely present.
2. All of this atmosphere escaped because of the high surface temperature of the molten, forming Earth.

B. After about 0.5 billion years, the crust started to form as Earth cooled.
1. Intense volcanic activity dominated the Earth's surface.
2. Today, volcanoes outgas 95–97% H_2O, 1–2% CO_2, and 1.5–2.5% SO_2, as well as nitrogen, chlorine, sulfur, and so forth.
3. Further cooling meant H_2O rained out of the atmosphere to form oceans.
4. Oceans then absorbed CO_2 through geological and biological processes starting 3.8 billion years ago, forming limestone.

- **C.** The first O_2 was produced by ultraviolet photons breaking up H_2O molecules in the atmosphere.
    1. As soon as O_2 gets close to 1%, ultraviolet interactions can produce O_3 (ozone).
    2. O_3 is very efficient at absorbing ultraviolet light, so no more is available to dissociate H_2O.
- **D.** Life, in the form of cyanobacteria, began producing lots of O_2 more than 3.3 billion years ago.
    1. Atmospheric abundance of O_2 remained very low as iron dissolved in sea water gobbled up the O_2 to make Fe_2O_3 (i.e., the Earth was rusting).
    2. "Banded iron formations," the main source of iron ore, are a testament to this process, which went on from 3.3 to 2 billion years ago.
    3. Another geologic feature, "red beds," testifies to a growing atmospheric abundance of O_2 starting 2.3 billion years ago.
    4. By 2 billion years ago, atmospheric O_2 had risen to more than 1% its current level, making eukaryotic metabolism possible.
    5. By 1 billion years ago, the rocks had saturated their thirst for O_2 and the atmospheric fraction started to rise.
    6. More atmospheric O_2 again meant more O_3, which protected surface-dwelling organisms from solar ultraviolet radiation, allowing the emergence of life on land.
    7. At 0.5 billion years ago, the Cambrian explosion brought a flowering of life; plants covering the land meant a huge release of O_2 into the atmosphere.
    8. In its entire history, only about 5% of the O_2 released into the atmosphere (30,000 trillion tons in total) is still there; the rest is tied up in minerals.
- **E.** The O_2 content has fluctuated between 15 and 35% over the past 0.5 billion years, with major consequences.
    1. Oxygen reached an all-time maximum between 320 million and 260 million years ago.
    2. Dragonfly wingspans reached 30 inches as their ability to absorb oxygen through their chitinous shell was enhanced by the greater abundance of O_2 in the air.

3. Forest fires were rampant—even wet material burns in a 35% O_2 atmosphere, and every lightening strike started an enormous forest fire.
4. Then, around 250 million years ago, CO_2 skyrocketed. There was a mass extinction in which about 90% of all species died, and O_2 fell by nearly $\frac{2}{3}$ to 12–15%; the cause(s) of these dramatic changes are much debated.

Suggested Reading:
Walker, *An Ocean of Air*.

Questions to Consider:
1. What is different about human modifications of Earth's atmospheric composition today compared to the much larger changes that have occurred throughout geologic history?
2. To what extent is the Earth a single system involving both biological and physical processes that will inevitably come into equilibrium such that life survives?

Lecture Fifteen—Transcript
The History of Earth's Atmosphere

The composition of the Earth's atmosphere has been heavily influenced by the evolution of life on Earth's surface, the topic of our last lecture. As a prelude to the development of a detailed record of our solar system's history, we'll focus this time on why planetary atmospheres in general, and Earth's atmosphere in particular, have the compositions that they do, exploring the forces that shape the atomic makeup of air.

The composition of a planet's atmosphere is determined by 3 factors: the strength of gravity on the surface of the planet, the temperature of the planet's surface, and the raw materials available to make molecules.

We should begin by asking the simple question: Why doesn't the atmosphere fall down? If you have a good memory, you'll remember the answer from my second lecture. The atmosphere is made of atoms; atoms have mass; the planet's gravity attracts everything that has mass toward its center. So why aren't all the air atoms lying on the ground? The answer is, of course, that air molecules do fall down. They are attracted to the planet, as are all other objects possessing mass. However, the thermal energy the atoms have, the kinetic energy of motion, the jiggling around which gives them what we characterize as a temperature, bumping into each other because they're warm, keeps the atmosphere up.

All atoms don't have the same speed because they have different masses, but for a given temperature, both an average speed and a distribution of speeds can be calculated. The average speed of an air molecule at room temperature is about 450 meters per second. At that speed, a typical nitrogen molecule traveling straight up and not bumping into any of its compatriots—an unlikely event—would reach 10,000 meters, about the height of Mount Everest, before falling back to Earth.

Since the temperature determines the average speed and the gravity determines the attractive force, these 2 quantities are critical for determining the thickness of the atmosphere. On Earth, it's remarkably thin. You may think, if you've ever climbed Mount Everest, that it's a very long way up and the atmosphere is thick, but to put it in perspective, if the Earth is represented by this orange, then the thickness of the atmosphere is less than this sheet of tissue paper. It's a very thin enveloping layer of gas that exists around our planet.

These 2 quantities, the temperature and the surface gravity, are also critical for determining the atmosphere's composition. To understand this, we must begin by dismissing an old saying that everything that goes up must come down. That's simply false. If that were true, how could we send satellites to other planets and even out of the solar system? We throw them up hard enough, and they don't come down.

Everybody, it turns out, has an escape velocity which, if you exceed it, you'll leave forever and not fall down. For example, if I take this orange and I throw it up in the air, throwing it gently means it comes back quickly. If I throw it harder, it takes a little longer to come back. But if I throw it hard enough, it doesn't come back at all. The orange has reached escape velocity and left the planet.

For the Earth, the escape velocity is determined only by its mass and its radius—that is, the mass and radius of the Earth—and is about 11 kilometers per second. That's pretty fast, 6 miles in a second, but not outside the range of the speeds which air molecules in the atmosphere can reach. Thus, any molecule in the atmosphere that exceeds this speed will escape from the Earth.

The speed of a molecule, as we've repeatedly encountered, is dependent on its mass. The light one moves fast and the heavy one moves slowly. Since nitrogen molecules have a mean speed of about 0.45 kilometers per second, more than 20 times lower than the escape speed from Earth, virtually none of them ever escape. However, hydrogen atoms—or even hydrogen molecules, 2 hydrogens joined together—have a mean speed of about 1.7 kilometers per second, so those moving only 6 or 7 times the average speed will escape.

You will recall that atoms colliding with each other have a distribution of speeds. And while the average is well determined by an equation relating the temperature and the mass, some molecules are moving much faster than that, so 6 or 7 times the average is not unusual at all. Indeed, all of the hydrogen originally found on Earth has escaped. Thus, among the solar system—which formed out of material that was 98% hydrogen and helium—virtually none of those atoms are left in Earth's atmosphere today.

Mercury, which is a small planet—which means it has a low mass, a relatively small radius, and therefore a low gravity—and is very close to the Sun—which means it's very hot, so any atoms in its atmosphere would be moving rapidly—in fact, has no atmosphere at all. Everything escapes. Not only is hydrogen and helium going fast enough to escape, as it is from the Earth, but the greater temperature, meaning a greater speed for things like nitrogen, oxygen, neon, and other gases, also are moving so fast that they can easily escape the tenuous gravity of Mercury.

Jupiter, on the other hand—which is huge, has an enormous amount of mass and therefore a very high gravitational force on its surface, and is far from the Sun, meaning it receives less solar energy and is therefore much colder—is composed mostly of hydrogen and helium. None of its original composition escaped, and therefore, its composition more closely resembles the Sun, that is, the cloud of material out of which all of the solar system formed.

The composition of Earth's atmosphere today, then, is a consequence of these fundamental forces which shape all atmospheres, as well as Earth's particular history. So what is, in detail, the composition of the atmosphere today? If I take a thimble-full of air, I contain millions of samples of a million molecules each. After all, atoms are very small. Let's take a million atoms and sort them by type. I have machines that do this; it's not particularly difficult.

Nitrogen and oxygen molecules—N_2, 2 nitrogens joined together, and O_2, 2 oxygens joined together—make up more than 98.7% of the atmosphere. In my million-atom sample, 778,888, give or take a few, atoms will be nitrogen, and 208,737, again, give or take a few, will be oxygen; more than 98.7% of all the air in this room. The third most abundant atom in the atmosphere might surprise you. It's argon, which is a noble gas. It's moderately abundant in the universe as a whole and it's also fairly heavy, so none of it has escaped from Earth. About 9300 out of every million atoms in the air in this room are argon. Since argon is a noble gas, it's inert; it performs no chemical reactions. You breathe it in, and you breathe it out, and nothing happens at all. It's just a constant exchange.

The fourth most abundant molecule in the Earth's atmosphere is water vapor. Roughly 2500 out of every million atoms and molecules will be water. I say roughly, because of course, the amount of water vapor in the air changes from day to day. You're well aware of that.

Sometimes it feels very sticky—that's humid—that means there are a lot of water molecules in the air, and some days it feels very dry—that's when there are fewer water molecules in the air that day. Averaged over the Earth it doesn't change very much, but from place to place, it certainly does. Anyway, roughly 2500 out of a million molecules of air will be water vapor.

Next is carbon dioxide, a critical molecule, as we'll see. Carbon dioxide, a C plus 2 Os, makes up about, today, 384 out of every million atoms in the atmosphere. This is a number that has changed drastically over time. We saw this in exploring the ice cores and in measuring the increase of carbon dioxide in the atmosphere as a consequence of human activity. Water vapor and carbon dioxide are 2 very important molecules because of the way they interact with the light from the Sun and the Earth, a subject I'll get to in a moment, when I finish my inventory.

Next on the list is neon. Only about 18 out of every million atoms are neon. Neon, of course, is another noble gas; it doesn't involve chemical reactions. We use it for making pretty colored signs.

And next is helium. Five out of every million atoms is helium. Now, I said that all the helium atoms originally present—given that helium is so low in mass, therefore has such a high speed, and therefore exceeds the escape velocity—are gone. The only helium atoms we have on Earth today are those that rain in from space with the interplanetary dust we've discussed before—a very small number of them—and the helium nuclei that are produced in radioactive decay of heavy elements, alpha decay, which are spit out as positively charged nuclei but soon accumulate a couple of electrons and become perfectly happy helium atoms filling our party balloons. Five out of a million today are helium. Now, the ones in the atmosphere have a high probability of escaping because they're light; like the ping-pong balls on the pool table, when you break with a pool ball, the ping-pong balls fly away. So the helium is constantly leaving the Earth. It's replenished by the continuing radioactive decay of the Earth's crust. However, since that's a 1-way street—once an atom undergoes an alpha decay, it's done, or it might undergo further decays, but that alpha is gone—then when that helium escapes, it has left the Earth. And as the radioactive decay slowly declines over the history of the Earth, the amount of helium in the atmosphere will decline, as well.

The next most abundant—which is not very abundant by this point—molecule in the atmosphere is methane, CH_4; 1.8 out of every million molecules of air are methane, but it also plays a critical role, along with water and carbon dioxide, in the greenhouse effect we'll discuss in a minute.

Next on the list is krypton; yes, there really is a krypton, although probably not kryptonite. One out of every million atoms is krypton, and it's just another of those noble gases that hangs out in the atmosphere. Being heavy, it can't leave.

Hydrogen makes up about 1 out of every 2 million molecules of air. Now, again, hydrogen left free in the atmosphere will rapidly escape, but hydrogen is highly reactive. So when, for example, a water molecule is split by an ultraviolet photon, a hydrogen atom might temporarily be present until it joins up with something else or leaves entirely. But 1 out of every 2 million in the steady state is hydrogen.

Then about 1 out of every 3 million are molecules that are mostly due to human activity, called nitrous oxides, and 1 out of every 10 billion or so molecules is another human-produced substance called chlorofluorocarbon, something that has been implicated in the destruction of the ozone layer and something in which we're trying to get control over our production.

Of these molecules, then, 98.7% of the atmosphere is nitrogen and oxygen, and another 0.93%, almost 1%, is made of the noble gases helium, neon, argon, and krypton, which never participate in any chemical reactions. The third most abundant molecule, then, that does participate in chemistry is water, and that contributes about $\frac{1}{4}$%, on average.

Carbon dioxide, methane, nitrous oxides, and chlorofluorocarbons make up the remaining 0.04%. But despite their very tiny fractions, they, along with water vapor, control the Earth's temperature through the greenhouse effect. It's not just the temperature that controls the composition of the atmosphere, but the composition of the atmosphere also controls the temperature.

The greenhouse effect works like this: Radiation leaves the Sun, characteristic of the speeds with which the atoms, or actually ions, in the outer atmosphere of the Sun are moving. The Sun's temperature is 5800°, which means those molecules and atoms and ions are moving very rapidly. The rapid acceleration of charges leads to rapid oscillations of electric fields, which propagate through space as light. Our atmosphere, fortunately for us, is completely transparent to this light. The light comes from the Sun right through the air and warms the surface of the Earth. Some of it actually is reflected away by clouds, for example, or upon reaching the Earth's surface, by other shiny things like glaciers—you all know the glare you have to wear sunglasses for when you're on a glacier or skiing in the wintertime. Those photons of visible light bounce right off and back into space and don't participate in warming the planet at all. Oceans are also somewhat reflective, water bodies, and deserts are more reflective than tropical jungles, of course, because tropical jungles are designed, with plants, to absorb photons from the Sun and produce organic matter. That's what plants do. They eat sunlight for energy.

So this energy coming in from the Sun, about 70% of it is absorbed by the ocean, by the ground, by the plants, by you lying on the beach, and about 30% of it is reflected back into space and doesn't participate any more in the energy balance which governs the Earth's temperature.

You might think that's the end of the story: Energy comes in and warms the Earth. But if that were true, it would be like an oven without a thermostat: You turn it on and it gets hotter and hotter and hotter, until you burn the house down. If energy is flowing into the Earth to keep a roughly constant temperature on the surface, then energy also must be flowing out from the Earth. Indeed, an equal amount of energy must be flowing out from the Earth. However, since the molecules in the Earth's surface—in the ground, in the ocean—are jiggling at a temperature characteristic of Earth, more like 300° above absolute zero or 20° on the centigrade scale, then they give off a wavelength, because of their slower accelerations, that's much more leisurely than the wavelength that the Sun puts out. Rather than visible light, they put out infrared radiation.

While the atmosphere is transparent to the high-energy photons coming in from the Sun, it is largely opaque to the infrared photons lazily making their way up from the Earth. In particular, carbon dioxide, water vapor, methane, nitrous oxides, and chlorofluorocarbons all have modes of oscillation that I talked about before—those vibrations between the atoms in the molecule and the tumbling motion that molecules make—that are very good at absorbing, sucking up, eating infrared radiation and exciting these vibrations and rotations. Thus, while infrared rays try to leave the surface of the Earth, they often get gobbled up on the way out and redirected back towards the Earth, acting like a blanket.

A blanket in your bed, in fact, works exactly the same way. As you lie there at night in the winter, you're cold. The reason you're cold is because the atoms in your body are jiggling back and forth and they're giving off infrared radiation at the rate of about the same as that of 100-watt light bulb. You radiate 100 watts. That energy, when it leaves you, slows down the molecules of your body, and that's what makes you feel cold. So what do you do? You get up and you pull on a blanket. What does the blanket do? As those infrared photons try to leave, the blanket absorbs them, turns them around, and redirects them back to you, and you develop this nice little warm cocoon around you.

Exactly the same thing happens with these so-called greenhouse gas molecules on Earth. The energy comes in from the Sun, but it can't escape, and so the temperature of the Earth warms up. Without these molecules, the Earth would be a very inhospitable place. Often in discussions today about global warming, the greenhouse effect is portrayed as an evil thing. In fact, the greenhouse effect is responsible for life on Earth. Without these infrared-absorbing greenhouse gases, the temperature of the Earth would be about 15° below zero on the centigrade scale. Everything on Earth would be frozen. And indeed, in fact, as we shall see, that has happened in the past. But because of this blanketing effect of the greenhouse gases, the temperature is raised to the point where water is in a liquid state and life can live comfortably on the surface of the Earth.

But while pulling on 1 blanket at night in the winter feels comfortable, pulling on 6 blankets soon won't, because then all of the infrared radiation trying to leave gets trapped, you get hotter and hotter and hotter, you start to sweat, and you kick the blankets onto the floor. So while the greenhouse effect is a good thing, adding more and more of these molecules to the atmosphere—carbon dioxide, and methane, and nitrous oxides, all as a consequence of human activity—is like piling on too many blankets and may lead to the point where the temperature rises too high.

It's unsurprising that most of the molecules in our atmosphere contain hydrogen, helium, carbon, nitrogen, and oxygen, since these are the 5 most abundant elements in the universe, and they constitute

99.6% of the mass of the cloud out of which the Earth was formed. The other planets all have atmospheres which are dominated by these same atoms, although Earth is unique in containing a substantial fraction of free oxygen.

The other planets' atmospheres are also determined by this balance of forces, the gravity which holds them in and the temperature which says how rapidly the atoms in the atmosphere are moving. And the temperature in these other planets, as well, is modified by the greenhouse effect. In particular, Venus has an extreme case of piling on blankets in that most of its atmosphere is carbon dioxide and the temperature on its surface is now 900°, a fate we certainly don't want to match.

The oxygen in our atmosphere, however, is what makes our atmosphere unique. Oxygen is highly reactive. The process of burning is combining with oxygen—that's pretty reactive—and the process of rusting is combining with oxygen, as well. Other atoms and molecules combine vigorously with oxygen. The continued and unique presence in our atmosphere of oxygen is a consequence of life.

The initial composition of the Earth's atmosphere was very different from that in evidence today. The initial atmosphere was almost certainly composed of the primordial elements, hydrogen and helium. In addition, the compounds of the most abundant other elements with hydrogen—like nitrogen plus 3 hydrogens, NH_3, or ammonia, and carbon plus 4 hydrogens, CH_4, methane—were likely also present. All of this atmosphere escaped because the high temperature of the molten forming Earth was such that the molecules moved so fast, Earth's gravity could not hold it in.

After about half a billion years, the crust started to form on the surface as the Earth cooled, and intense volcanic activity dominated everything on Earth. Today, we can measure the gas that flies out of volcanoes when they erupt. It tends to be about 95 to 97% water vapor, 1 to 2% carbon dioxide, 1.5 to 2% sulfur dioxide, as well as nitrogen, chlorine, sulfur, and other compounds. Further cooling of the Earth meant that the water vapor that came out of volcanoes eventually wasn't moving fast enough to escape, and as it rose in the atmosphere and cooled—that is, slowed down—little molecules of water got together to make little drops of water, and when they got heavy enough, they rained out to form the oceans.

The oceans then absorbed a lot of the carbon dioxide through geological and biological processes, starting about 3.8 billion years ago when, as we saw, the first bacteria lived. Geological processes take carbon dioxide out of the water, as well, and make limestone.

The first free oxygen, O_2, oxygen molecules, were actually produced not by light, but by ultraviolet photons from the Sun breaking up the water molecules that came out of volcanoes. That is, you have an oxygen and 2 hydrogens; if it absorbs an ultraviolet photon, the bond that holds the hydrogen to the oxygen vibrates so violently that the hydrogen just disappears. Knock off the other hydrogen, as well, and you have a free oxygen atom. These free oxygen atoms are very reactive, as I've said, and they quickly find another oxygen atom to join up with, making O_2. So the first oxygen came from a nonbiological process, the ultraviolet radiation from the Sun.

Ironically, however, as O_2 builds up in the atmosphere and gets close to 1%, as opposed to the 20% it is today, it actually starts absorbing ultraviolet radiation, breaking it in half—the O_2 into 2 single oxygens—and then, with a bunch of other O_2s present for a single oxygen to find, joining together in a triplet of oxygen atoms, O_3, which is called ozone.

It turns out ozone is bound in such a way that it can readily absorb ultraviolet light without breaking. Indeed, the ozone layer above the Earth is very efficient at absorbing the ultraviolet light from the Sun, meaning there was no more ultraviolet light available to disassociate the water coming out of the volcanoes, and the amount of oxygen in the atmosphere stabilized at a low level of around a percent.

Life in the form of cyanobacteria, that we met last time, began producing lots of O_2 roughly 3.5 to 3.3 billion years ago. The atmospheric abundance of oxygen, however, remained relatively low. As we have learned, iron dissolved in the seawater gobbled up the O_2 to make Fe_2O_3, iron oxide. A way to say this is: The Earth was rusting. The banded-iron formations I discussed last time, the main sources of iron ore that we use today, are testament to this process, which went on for well over a billion years. Another geological feature, called red beds, testifies to the growing abundance of atmospheric oxygen; as the iron eventually got saturated, other atoms and molecules started to join with the O_2, starting about 2.5 billion years ago.

By 2 billion years, atmospheric oxygen had begun to rise over 1% of its current level, making it possible for the new kind of life then emerging: the eukaryotes, the cells that contained a nucleus and, riding along for the ride, little symbiotic mitochondria inside the cell which can use oxygen from the air in providing energy for the cell's life. By 1 billion years ago, the rocks and the iron dissolved in the ocean had saturated their thirst for oxygen in the atmosphere, and the atmospheric fraction, therefore, started to rise.

More atmospheric oxygen again meant more ozone, which provided an even thicker ultraviolet shield from the Sun's radiation and, therefore, protected surface-dwelling organisms in the ocean and allowed the emergence of life on land, protected from this high-energy radiation.

By about a half a billion years ago, the Cambrian explosion brought a flowering of life; plants covering the land meant a huge release of oxygen to the atmosphere. In its entire history, only about 5% of the oxygen released into the atmosphere, a total of about 30,000 trillion tons, is still there. Most of the rest is tied up in minerals.

The O_2 content has fluctuated between about 15% and about 35% over the last half billion years, with major consequences. The oxygen reached an all-time high between about 320 and 260 million years ago, sort of the dawn of the age of the dinosaurs. The creatures that could survive in this very oxygen-rich atmosphere were really quite extreme. Fossils show dragonflies with wingspans of 30 inches. Today's insects in the atmosphere are limited in size by the amount of oxygen, which they need for life, that can be absorbed through their chitinous shells. They don't have lungs per se but absorb oxygen directly through their shells. And the surface area to volume of the insect—the volume determining the amount of oxygen needed and the surface area saying the amount of oxygen that can be absorbed—sets a maximum size on insects. As the oxygen level rises, the insects can get larger and larger because it's easier and easier to get oxygen through the shell, and dragonflies with wingspans like this would be quite startling to see.

Another consequence of this high, 35%, fraction of oxygen, however, is that forest fires were rampant. Even wet material burns in a 35% oxygen atmosphere, and every lighting strike started enormous forest fires around the world. Then, about 251 million years ago, the carbon dioxide in the atmosphere skyrocketed, there was a mass extinction which wiped out 90% of all species, as I said earlier, and the oxygen in the air fell nearly $\frac{2}{3}$, to the range of 12 to 15%. The cause of these sudden and dramatic changes is still a matter of much debate.

The complex history of an ever-changing atmosphere shows what we can learn about 1 planet in 1 solar system—our planet. Now we are about to take a step back and take a much broader view of our place in the universe, which will require a vast increase in our scales of space and time. We'll begin this quest by exploring the origin of the solar system and determining a precise date for its birth.

Lecture Sixteen
The Age of the Solar System

Scope:

Remarkably, we now know the age of our solar system to an accuracy of better than 1%: 4.56 billion years. This date comes not from Earth rocks, which are in a state of constant flux on our planet's dynamic surface, but from isotope ratios of rubidium and strontium in the oldest meteorites. From these and from rocks brought back by the Apollo astronauts, we can not only date the solar system's birth but can infer the sequence of events that formed the planets and even understand the origin of our unusually large moon as the consequence of a massive collision near the end of the Earth's formation.

Outline

I. Assertions of the Earth's age come from many sources.
   A. The Babylonians and the Greeks assumed the universe (and the Earth) were infinitely old, obviating the question.
   B. The earliest known quantitative estimate is a Hindu chronology putting the age at precisely 1,972,949,091 years in 150 B.C.; in fact, this is closer to the true value than any subsequent estimate prior to the middle of the 18th century.
   C. Monotheistic religions, which require a moment of creation, made much shorter estimates.
      1. Zoroaster estimated 12,000 years in 1400 B.C.
      2. Hebrew, Christian, and Islamic scholars made the Earth younger still.
      3. The official biblically based estimate for the Earth's creation date is from Bishop Usher in 1650: 5 p.m., October 23, 4004 B.C.
   D. With the rise of science in the 17th century, new methods were brought to bear on the question.
      1. Kepler made the first "scientific" estimate of Earth's age based on the change in the apogee of the Sun, but, influenced by the philosophy of the time, his answer was 5993 years.

2. Benoit de Maillet, noting the prevalence of fossil seashells on land and therefore assuming the Earth was once covered by water, concluded in 1748 that the Earth was at least 2 billion years old, based on his measurements of the decline of sea level (which was bogus).
3. In the mid-19th century, physicists split with geologists and biologists: Lord Kelvin, arguing from the cooling rate of the Earth and lifetime of the Sun, estimated about 100 million years, while those believing their observations required slow geological and biological evolution argued for at least 2 billion years.
4. The astronomer Henry Norris Russell first used radioactivity (uranium → lead) to estimate 4 billion years as the Earth's age; since then, refinements in this technique have converged on today's answer.

II. Radioactive "accumulation clocks" are read by simply looking at the ratio of parent to daughter isotopes in a sample.
   A. The total of parent plus daughter nuclei gives the amount of the parent originally present.
      1. Knowing the half-life and the fraction that have decayed gives the age directly.
      2. This must assume no daughter was present at the start, and that no daughter nuclei have escaped from the sample over its lifetime.
   B. A good example of an accumulation clock is potassium-40.
      1. It undergoes a beta decay to argon-40 89% of the time; otherwise it decays to calcium-40.
      2. The half-life is 1.25 billion years.
      3. Since argon is an inert gas, it undergoes no chemical reactions and stays trapped until the rock containing the original potassium melts.
      4. This makes it a very useful tool for relative dating of rocks on the Earth and Moon but poor for a determination of the solar system's age, since all rocks were originally molten.
   C. Analysis of Moon rocks brought back by the Apollo astronauts reveal its origin.
      1. The ratio of the mass of the Moon to the mass of the Earth is unique in the solar system—most moons are much smaller compared to their parent planets.

2. Dating Moon rocks with the samarium-neodymium decay sequence shows the oldest are 4.46 billion years old, representing a solidification less than 100 million years after the Earth's formation.
3. All of the Moon rocks are characteristic of rock found in the Earth's crust, rather than being a mix of primordial solar system material.
4. Our conclusion is that the Moon formed when a giant planetesimal nearly the size of Mars collided with Earth as it was completing its formation and sprayed off enough material to produce a planet-sized moon.

D. Uranium-235, uranium-238, and thorium-232 are also good accumulation clocks because they each decay to a unique isotope of lead.
1. This takes many radioactive decay steps; e.g., uranium-238 takes 14 steps (6 beta decays and 8 alpha decays) to reach lead-206.
2. The intermediate steps are relatively fast, such that "secular equilibrium" is set up: The number of nuclei of a given isotope is equal to the lifetime of that isotope.

E. The problem with all such clocks on Earth is that tectonic activity leaves no rocks on the surface as old as the Earth itself.

III. The oldest unchanged rocks in the solar system are asteroids, interplanetary rocks that never formed into a planet.
A. Chunks of asteroids fall to Earth as meteorites and can be used to date the origin of the solar system.
B. The strontium-rubidium clock yields the most precise measurements.
1. Rubidium-87 constitutes 30% of naturally occurring rubidium and decays to strontium-87 via beta decay, with a half-life of 48.8 billion years—4 times the age of the universe, so there is plenty still around.
2. This cannot be used as a simple accumulation clock, since the amount of strontium-87 present to start with is unknown.

3. This problem is cleverly circumvented by establishing "isochrones," lines derived from taking the ratio of both strontium-87 and rubidium-87 to the stable, nonradioactive product strontium-86.
4. Measurements taken from more than 2 dozen meteors all gave results consistent with an age of 4.56 ± 0.025 billion years, accurate to 0.5%.

Suggested Reading:

Dalrymple, *The Age of the Earth.*

Questions to Consider:

1. What arguments might a young-Earth creationist use to refute the age of the planet as derived from measurements of radioactive isotopes?
2. If the body that struck the Earth and created the Moon had just missed us, what would be different today?

Lecture Sixteen—Transcript
The Age of the Solar System

The rich history of our atmosphere, oceans, landforms, and life that we have deduced from our atomic historians had a finite beginning. More than 9 billion years after the universe came into existence, a smallish cloud of interstellar gas and dust coalesced to form a sun, and the detritus left over from this event produced the retinue of planets that orbit it today.

It's important to note that this just didn't happen once in the history of our galaxy. Until 1995, we didn't know whether or not our solar system was unique. We knew there were many other stars like the Sun, but we had no clue as to whether or not these stars also hosted planets. Now we do know—many of them have planets. In the last 15 years, over 330 other solar systems have been discovered, and a new one is now found almost once a week.

It's not a trivial thing to uncover planets orbiting a distant star. Recall from Lecture One, how if the Sun is represented by an orange, the Earth was just a grain of sand about 15 feet away, and the next orange, the next star, was in Minneapolis, with a little grain of sand orbiting it just 15 feet from it. Taking a direct picture is extremely difficult because the star is so bright and the planet so dim. And so we've reverted to indirect methods to attempt to see whether our solar system was unique in the universe.

One of these is to take advantage of the slight tug from gravity that a planet exerts on its parent star. One might imagine that the star sits still in the center of a solar system and the planets orbit around it. That's the picture one gets in a textbook. However, the 2 objects both have mass and each attracts the other. And so when the planet is over here it tugs on the sun, and when the planet is over here, it tugs the opposite direction on the sun. And while the planet makes a big circle, the sun makes a tiny little circle. That tiny little circle amounts to a velocity of only a few meters per second, but using careful observations of the spectral lines from the atmosphere of the star, we can actually detect these very modest motions and, in this way, have discovered most of the several hundred other planetary systems.

In certain special cases, the planet orbiting its star is aligned almost perfectly with our view such that when it passes in front of the star, it blocks a little bit of the star's light—typically much less than 1%, so one needs very precise measurements of the light from the star, but that eclipse allows us to measure not only the time it takes the planet to go around, but the diameter of the planet as it moves across the star and other parameters of its structure. Several dozen of these eclipsing planet-star systems have already been discovered.

There are biases in our observations of these other solar systems. It's clearly easier to detect very large planets, which have a big pull on their parent star and/or, when they pass in front of them, block a significant fraction of light. Most of the planets we've discovered to date are the size of Jupiter, give or take a little bigger or a little smaller. A few are the size of Neptune, but that's still massive compared to the Earth. In the coming year however, 2009, NASA will launch a satellite named after the astronomer Kepler, which will be able to actually detect eclipses of Earth-like planets around Sun-like stars. We will then know for certain how common planets like ours are.

However, even if planets are common, it would be nice to know more about our own and how it came to be. Speculating about the birth of our special case, the age of the Earth itself, is a pursuit that's probably as old as the human species. Today, thanks to a pair of rare atoms, we know the answer quite precisely: 4.56 billion years.

Assertions as to the age of the Earth have come from many sources. The Babylonians and the Greeks both assumed the universe and, therefore, the Earth were infinitely old, obviating the question. If something's been around forever, its age is a meaningless concept. The earliest known quantitative estimate for a finite age of the Earth is found in Hindu chronology, and it put the age very precisely at 1,972,949,091 years in 150 B.C. Curiously, in fact, this is closer to the true value than any subsequent estimate for the next 1700 years. I don't impute any special meaning to this approximation by the Hindu chronologers, except to say perhaps that they had a better perspective on their significance in the universe than Western philosophers seem to have done.

In monotheistic religions one requires, almost by definition, a creator because one has an all-powerful creator. Most of these religions made much shorter estimates for the age of the Earth. Zoroaster in ancient Persia estimated the Earth to be 12,000 years old in about 1400 B.C. Hebrew, Christian, and Islamic scholars made the Earth younger still. Theophilus in 200 A.D. estimated 7531 years, rather precise. St. Basil in 350 A.D., 6006 years. St. Augustine, 100 years later, at 6333 years. The level of precision of these estimates lies outside the ethos of modern science, where we only quote numbers to the number of figures to which we believe we have the answer. In other words, we leave some uncertainty. But these biblical estimates, of course, were based on summing up the "begats" in Genesis and the ages of each of the ancients whose lives were reported there.

The official biblical estimate that has stood for a long period of time for the Earth's creation date is from Bishop Ussher in 1650. James Ussher was the Anglican bishop of Armagh in Ireland and the primate of all Ireland for the Anglican Church. And, over a long scholarly period, he made the most precise estimate for the day of creation: It was at nightfall on the day proceeding October 23, 4004 B.C.

With the rise of science in the 17th century, new methods were brought to bear on the question of the age of the Earth. The astronomer Kepler actually made the first "scientific" estimate, where scientific is in quotes here, based on the change of the apogee, or maximum distance, of the Earth and Sun in the Earth's orbit around the Sun. But, of course, as we must always be aware, scientists even can be influenced by the philosophy of their time, and his answer was 5993 years, not very far from Bishop Ussher's. Both the data on which he based this estimate and the model he used to compute it were wrong, and so was the date.

Benoît de Maillet, a French natural philosopher in the 18th century, noted that there was a prevalence of fossils of seashells on land and, therefore, not unreasonably, assumed that the Earth was once covered by water. He made regular annual measurements outside his house near the sea and concluded that the sea level was gradually going down. Calculating from how high the sea must have been to cover the Alps and given that it was receding at this very slow rate, he estimated that the Earth was at least 2 billion years old. Now, his measurements of the decline of the sea were bogus, of course, because the sea was in fact not shrinking, at least to the levels that he could measure it, but nonetheless, we finally got back to a number measured in billions of years.

In the 1800s, other methods, presumably scientific, were also applied. The accumulation of salt in the sea, for example, washing from the land, gave an estimate also of about a billion years. And the tidal effects of the Moon on the Earth were viewed as another method of calculating the age of the Earth.

In fact, there's an interesting note here that the Moon is affecting the Earth and affecting its rotation. As the Moon moves around and the Earth spins on its axis, as you know, a tide is raised each day, twice a day, as the water rises in the direction of the Moon and falls 90° away. This actually creates a friction on the Earth which is slowing its rotation. Simultaneously, the Moon is moving farther from the Earth to compensate for that slowing of the rotation. The Moon used to be much closer to the Earth and the Earth used to spin in much less than 24 hours. Several billion years from now, the Earth and the Moon will reach a state where they are locked together such that the face of one always faces the other. The day will then become 47 days long—that is, it'll take 47 days for the Earth to rotate once on its axis—and the Moon will always be over the same hemisphere of the Earth. So it'll give a new definition to the idea of a honeymoon. To go see the Moon, you may have to take a trip halfway around the world. It'll become a new tourist destination, whichever side of the world the Moon ends up on. Nonetheless, though this tidal effect is important, it doesn't allow us an obvious age for the Earth.

In the mid-19th century, an enormous debate erupted between physicists, on the one hand, and geologists and biologists, on the other. Lord Kelvin, who was certainly the leading physicist of the Victorian age, argued from the cooling rate of the Earth that it could not be more than about 100 million years old. The geologists and biologists, believing their observations required slow evolutionary processes, argued for an age of at least 2 billion years—back to that Hindu number.

Kelvin did his calculation in the following way: He assumed the Earth was formed hot and it had simply been cooling off since. This is largely a correct assumption. Any of you who have been down a deep mine know that it's warm down there. In fact, as we've seen, the center of the Earth is a temperature of nearly 5000°, and the metals there, iron and nickel and others, are molten. Indeed, the Earth is warmer as you go in, and it is cooling off, gradually radiating its energy to space. Kelvin had to figure out first what mechanism was involved in transporting heat from the center of the Earth to the outside.

There are 3 ways in which heat—which you will recall is just the energetic jiggling of atoms—can be moved from one place to another. The first is radiation. That's the way heat gets from the Sun to the Earth—electromagnetic waves of energy which carry energy away from one source to another.

A second method is convection. That's what you see when you boil a pot of water on the stove. The flames of the stove heat the bottom of the pot, which makes the atoms in the pot jiggle very rapidly. Those collide with the water molecules that are adjacent to the bottom of the pot, making them jiggle very rapidly and making them want to expand. And so, in bulk, the water at the bottom of the pot rises to the top of the pot and you get a convective turnover, a big motion of macroscopic quantities of matter because in one place it's hot and in another place it's cool.

The third mechanism of heat transport is what we call conduction, and that's the microscopic jiggling of atoms next to each other, pounding into each other and moving heat along. That's the effect you feel when you put a silver spoon into a cup of hot tea and touch the end of the spoon. It's very hot. That's because the tea, the water molecules, have been jiggling rapidly, pounding into the molecules of the silver, making them jiggle rapidly, and then bumping into their neighbors, transferring the heat up the spoon. Kelvin assumed that the latter, conduction, was the mechanism by which most of the energy came out of the Earth; again, a correct assumption.

There were then 3 more bits of information he needed. One was the initial temperature of the Earth, which he, not unreasonably, assumed was about 3900°, because that's the temperature in which laboratory experiments had shown that rocks become molten. Second was the gradient in temperature, the amount of change for every meter you went up from the center of the Earth to the edge. This could be actually directly measured in mines or deep wells but, of course, only in the top mile or 2 of the Earth. What the gradient was inside, he had to only speculate. He measured this gradient to be about 0.04° C for every meter traveled up from the center of the Earth.

Finally, he had to measure the conductivity of the rocks, the rate at which rocks would transfer these jiggling atoms and move the energy up. That he did in a laboratory by measuring rocks on the surface. But, of course, as we saw, surface rocks are not representative of the Earth as a whole. Surface rocks have a density of 2 to 3 grams per cubic centimeter, whereas the density of the center of the Earth is more like 10 grams per cubic centimeter, so that was another uncertainty in his measurement.

In any event, he did the calculation and came up with a number of 98 million years, although unlike some of the biblical accounts, he did give a range, which was extraordinarily large, from 20 million to 400 million years. He did miss a critical thing, of course, and that is that while the Earth was formed warm and is cooling off, it also is kept warm by the radioactive decay of atomic nuclei. Since atoms weren't firmly established and nuclei were certainly unknown, Kelvin is to be forgiven for missing this important contribution to the energy balance of the Earth. Nonetheless, being a physicist, he was quite confident of his prediction, and he wrote:

> It seems, therefore, on the whole most probable that the Sun has not illuminated the Earth for [more than] 100,000,000 years, and almost certainly ... has not done so for 500,000,000 years. [This is as some geologists had speculated.] As for the future, we may say, with equal certainty, that the inhabitants of the Earth cannot continue to enjoy the light and heat essential to their life for many millions of years longer, unless [and this is a critical qualification, unless] sources now unknown to us are prepared in the great storehouse of creation. [Lord Kelvin, 1862]

He stuck to this throughout his life. Thirty-five years later, he did a separate calculation to reinforce his notion that the solar system was less than 100 million years old, and that was to calculate the lifetime of the Sun. There were 3 possible mechanisms he could think of for the energy that the Sun constantly radiated into space.

One was that the Sun was literally shrinking, that when things fall down, they give off energy, and so if the Sun had started out very large—which we'll see in a future lecture that it did—and it was constantly shrinking, gravitational energy would be given off, and that could be given off in the form of light, which would illuminate and warm the Earth. A simple calculation shows that the amount of energy the Sun produces requires a tiny amount of shrinking of this massive body such that it would never have been observed in a single human lifetime or even throughout human history. So it wasn't a matter of being inconsistent with seeing the Sun going *sssssss*, like that. However, when you do the calculation and see how long it takes to shrink all the way to 0, the answer is only a few tens of millions of years, not inconsistent with his measurement from the cooling of the Earth.

The second idea he considered was that the Sun, like the Earth, was simply born hot and had been cooling off. If you measure the surface temperature of the Sun and assume it's that temperature all the way through, it's not hard to come up with an estimate for how long it would take for the Sun to radiate all that thermal energy, that jiggling of atoms, away. And, again, he comes up with a number that's in tens of millions of years.

The third alternative, the only other energy source he could think of, was some equivalent of burning, whether it was the combining of other elements with oxygen or some other chemical process, some chemical rearrangement of atoms in the Sun that would, as burning does, give off energy. Again, taking the typical amount of energy given off when 2 atoms or molecules come together and form a new one, one again comes up with an estimate of only a few tens of millions of years.

Taking all these 3 sources into account, the only ones he could imagine, in 1897, 35 years after his first estimate, Kelvin published another estimate of the lifetime of the solar system, based on how long the Sun could have been shining from each of these methods, and that was 30 million years. Throughout his life, then, he stuck to the notion that physics would only allow the solar system to be 100 million years old, despite the urgent need of evolutionary biologists and geological theory of the time that slow, uniform changes in the Earth would take billions of years.

In fact, it wasn't until 1921 that we had our first realistic estimate for the age of the Earth. The astronomer Henry Norris Russell, whom we will encounter in a future lecture, was the first to use radioactivity, discovered in the first decade of the 20th century, and in particular, the radioactive decay of uranium to lead, to estimate that the crust of the Earth must be about 4 billion years old. This, of course, was by far the best estimate to date, and since then, refinements in this technique have converged on today's answer.

Radioactive accumulation clocks, that I mentioned in Lecture Five, work very simply by looking at the ratio of the parent radioactive nucleus to the daughter isotopes that it produces. The total of the parent plus the daughter gives the amount of parent originally present in the sample, and then, knowing the half-life, the fraction that has decayed, gives the age directly. This procedure must assume 2 things, however: (1) that there was no daughter present at the start—otherwise that would mess up the ratio—and (2) that no daughter nuclei have escaped, once produced, from the sample over its entire lifetime.

A good example of an accumulation clock is potassium-40, which we've encountered earlier in the course. It undergoes a beta decay to argon-40 89% of the time. The rest of the time it decays to calcium-40. The half-life of this dominant potassium-to-argon decay is 1.25 billion years, a convenient number if we want to measure something like the life of the Earth, which we now think is billions of years old.

Since argon is an inert gas, one of the noble gases, it undergoes no chemical reactions of any kind, and it tends to stay trapped in the rock until the rock in some way melts. This makes it a very useful tool for the relative dating of rocks on Earth from the time they first solidified to the present and has also been used for rocks on the Moon. But it makes for a poor determination of the age of the solar system as a whole, since all the rocks were originally molten, and therefore, the original decay of potassium to argon would have missed some of the daughter nuclei, which would have escaped.

The analysis of Moon rocks brought back by the Apollo astronauts helps to reveal the origin of the Moon itself. The ratio of the mass of the Moon to the mass of the Earth is unique in the solar system. Most moons are much smaller compared to their parent planets. Mercury and Venus have no moons at all. Mars has 2 pathetic little moons, 12 and 22 kilometers across, respectively, not too much bigger than the asteroid that killed the dinosaurs, and therefore, have a much tinier ratio of the mass of the moon to the mass of the planet. Some of Jupiter's moons are comparable in size to our Moon, but Jupiter itself is 318 times the mass of the Earth. So again, the ratio of the mass of the moon to the mass of the planet is a much smaller number. The Earth and its Moon are unique in this regard.

We date the Moon rocks with a new radioactive sequence that we haven't met yet in the course, the samarium-to-neodymium decay. These 2 elements are joined in a parent-daughter relationship by an alpha decay from samarium-147 to neodymium-143. The half-life is an extraordinary 106 billion years, plenty long enough that there's a whole lot of samarium still around. This dating technique, a simple accumulation clock of comparing the ratio of daughters to parents, shows that the oldest rocks on the Moon, that is, the oldest rocks since they were last molten, are 4.46 billion years old; as we will see, only 100 million years after the Earth's formation itself.

All of the Moon rocks are characteristic of rocks in the Earth's crust, rather than even being a mix of the primordial material of the solar system. Recall that the Earth rocks are much lighter or less dense than the Earth is overall, but the Moon, the entire Moon, has a density similar to the crustal rocks on the Earth. The conclusion for this, which slowly built up over the last several decades, is that our Moon formed when a giant planetesimal, a giant proto-planetary chunk of material floating around in the solar system, that was nearly the size of Mars, collided with the Earth just as the Earth was completing its formation, spraying off enough material to produce what ultimately became a planet-sized moon.

Uranium-235, uranium-238, and thorium-232 are also good accumulation clocks, because each of them decays to a unique isotope of lead. This takes many radioactive decay steps, as I mentioned earlier. For example, uranium-238 takes 14 steps, 6 beta decays and 8 alpha decays, to get all the way down to lead-206. The intermediate steps are relatively fast, such that a secular equilibrium is set up. By "secular equilibrium," I mean that the number of nuclei of each of the intervening isotopes is equal to the lifetime of that isotope.

A simple analogy with visitors to a museum helps make this picture clear. In New York, at least, when there's a special show, say at the Metropolitan Museum, they don't just open the doors and let the hordes of people come in. You get a ticket that's stamped with a time that says you must show up between 9 o'clock and 10 o'clock, say, on Wednesday morning, if you want to see this special exhibit.

Let's say the museum hands out 100 tickets for each hour of the days that the special show is going to be opening, and the special show occupies 4 rooms in the museum. And let's further assume that the average visitor spends an hour in the first room, 2 hours in the second room because it's got more stuff, only 30 minutes in the third room because they're getting tired, and an hour in the last room which shows the ultimate treasures of the exhibit. If that's the case, and people enter at the rate of, say, 1 a minute in the first room of the exhibit when the museum opens in the morning, then in the first minute, there'll be 1 person in the first room, in the second minute there'll be 2, and then 3, and then 4, and then 5, until at 60 minutes there are 60 people in the first room. On the 61st minute, another person enters the museum, but having spent the average of an hour in that first room, the first person who entered moves onto the second room. There therefore still are 60 people in the first room, but now 1 in the second room, as well.

This process continues with people shuffling down the line, until after 3 hours, where people spend 2 hours in the second room, there are 60 people in the first room and 120 people in the second room. Three hours and 1 minute: the first person who entered the museum moves from the second to the third room, where he or she will spend only 30 minutes. And we continue this conveyor belt until we see we will have 60 people in the first room, 120 people in the second room, only 30 people in the third room because they only stay there for 30 minutes, and 60 people in the last room, before the first person who enters the museum leaves it $3\frac{1}{2}$ hours later.

What does this mean? This means that the number of people in any room in the museum is just proportional to the amount of time they spend in that room of the museum. That's what I mean by a secular equilibrium, and the same thing occurs in this radioactive sequence from uranium-238 to lead-206. The time an isotope spends in the state, that is, its half-life, determines how many of that isotope exist in the sample. And by measuring all 14 different isotopes in this chain, one can get a rather accurate estimate of the age of the rock.

The problem with all such accumulation clocks on Earth is that tectonic activity—volcanoes and plates moving around and colliding with each other and diving under the ocean—over the lifetime of the Earth means that there are no rocks on the surface of the Earth that are as old as the Earth itself. The oldest unchanged rocks in the solar system are asteroids, interplanetary rocks that never managed to form into a planet. Chunks of these fall to Earth occasionally as meteorites and can be used to date the origin of the solar system itself. Over 30,000 meteorites, from the size of tiny marbles to the size of beach balls, fall to Earth every year, although only about 1% are ever recovered.

The clock that has been most helpful in dating these original components of the solar system is the strontium-rubidium clock which I mentioned earlier. Rubidium-87 constitutes 30% of naturally occurring rubidium and decays to strontium-87 via a beta decay with a half-life of 48.8 billion years, 4 times the age of the universe, so there's still plenty of it around.

This cannot be used as a simple accumulation clock since the amount of strontium-87 present to start with is unknown. This problem, however, is cleverly circumvented by establishing a procedure using isochrones, lines derived from taking the ratio of both strontium-87 and rubidium-87, the parent and the daughter, to the stable nonradioactive product, strontium-86. When I first encountered this technique, it was a stimulation to look further into the subjects I've discussed in this course, because it is elegantly simple and yet remarkably clever.

What one does is make a graph, a plot, by measuring the ratio of strontium-87 to strontium-86 and rubidium-87 to strontium-86, plotting the first on the y-axis and the second on the x. Then, through some relatively straightforward algebra, which I won't trouble you with but will supply upon request—in this plot, (*1*) we measure the 2 ratios directly from the meteorites, so those are measurements that are unequivocal; (*2*) the amount of strontium-87 present at $t = 0$, the moment the meteorite first condensed out of the solar material, is given by the point on the y-axis where the line connecting the measurements of strontium-87 to strontium-86 crosses the y-axis, the so-called y-intercept; and the slope of the straight line that fits all these measurements gives the age directly.

These measurements have been done on over 2 dozen meteorites, and they all give results consistent with one another. The age of the Earth is 4.56 ± 0.025 billion years. That is accurate to better than $\frac{1}{2}$%. Now, this is the age of the solar system and, to a very good approximation, the Earth. Not 6000 years, not 2 billion years, not an infinite age, but 4.56 billion years, roughly $\frac{1}{3}$ the age of the universe.

This elegant technique provides the answer to a question we did not originally ask; it tells us the age of the meteorites. But how long did they take to form? Was the Earth's formation contemporaneous? And perhaps even more fundamentally, why did the Earth form? What triggered the formation of the whole solar system in the first place? I hope by now you won't be surprised to learn that our little atomic historians are up to the task of answering these questions, as well.

Lecture Seventeen
What Happened before the Sun Was Born?

Scope:

The atoms that comprise the bodies of the solar system existed long before the Earth, Sun, and Moon could form. They carry with them, in their isotopic ratios, signatures of earlier events. In particular, the unusually large amount of a particular isotope of magnesium in meteoritic material suggests a cataclysmic event in our vicinity just a few million years before the Sun formed: the explosive destruction of a massive star. Such an event, the only known source for creating the radioactive isotope aluminum-26, not only explains the magnesium anomaly but might also have provided the push that a nearby cloud of gas needed to collapse and form our Sun. As a bonus, the remnant of the exploded star could have supplied the source of special radiation needed to produce our exclusively left-handed form of life.

Outline

I. Traces of extinct radioisotopes, long vanished through decay, provide clues as to how long it took for solid material to condense out of the protosolar nebula.
   A. A solar system forms as a spinning cloud of interstellar dust and gas collapses into a sphere, which makes the star and a thin disk of orbiting detritus from which the planets form.
      1. We see this process at work in star-formation regions today.
      2. Different materials condense out into solids at different temperatures, and thus at different distances, from the forming star.
      3. The exact timescales for these processes to occur, however, are difficult to calculate from first principles.
   B. The radioactive isotope plutonium-244 provides a direct measure of the time it took this process to proceed.
      1. Plutonium-244 is produced in stars and has a half-life of 83 million years.
      2. It decays through 2 alpha decays to thorium-232.

3. Thorium-232 occasionally decays through the rare process of spontaneous fission, where the nucleus splits roughly in half, spitting out several extra neutrons. The half-life is 13.9 billion years—roughly the age of the universe.
4. Since hundreds of mega–electron volts are given off in a fission, the event leaves microscopic tracks in the meteoritic material in which the splitting atom was embedded.
5. In addition, the expelled neutrons combine with other nuclei to make exotic isotopes.

C. Both iodine-128 and iodine-129 are produced by neutron capture from the dominant stable isotope, iodine-127.
1. Both heavy isotopes of iodine decay via beta decay to the corresponding isotopes of xenon.
2. As a noble gas, xenon never combines with anything to make compounds and should not be present at all in condensed, rocky material.
3. When examining meteorites, one finds both xenon-129 and xenon-128, together with iodine-127 from which their now-deceased parents (iodine-129 and iodine-128, respectively) are made.
4. The ratio of xenon-129 to iodine-127, roughly 1 to 10,000, provides a clock to measure how fast the material condensed. A high ratio implies early formation, while a low ratio implies the material condensed later after most of the xenon had escaped.
5. The value for carbonaceous chondrites, the most primitive meteor type, show a very narrow spread in condensation times, around a value of 12 million years.
6. This provides us with a quantitative clue about the earliest events in the formation of the planets.

II. Extinct radioisotopes also provide a clue as to what happened just before the solar system formed.
A. Many metallic meteorites show obvious signs of differentiation.
1. The heavy metals are on the inside, and the lighter minerals are on the surface.
2. This provides strong evidence that after coalescing, the asteroids were in a molten state.
3. Their masses are much too small for gravitational heating to have melted the material, as happened on Earth.

- **B.** Radioactive decay is an alternative heating source that could explain the differentiation.
    1. Aluminum is one of the most abundant elements other than hydrogen and helium, making up more than 2% of all heavier elements.
    2. Aluminum-26 is a radioactive isotope that constitutes 50 atoms out of every million aluminum atoms.
    3. It undergoes beta decay with a half-life of only 730,000 years, releasing 4 MeV per decay.
    4. Since aluminum-26 makes up 1 in every million atoms of material and it takes much less than 4 eV to melt this material, this single isotope is enough to do it.
- **C.** The abundance of magnesium-26 in meteorites shows there was a lot of the short-lived aluminum-26 around when the solar system was formed.
- **D.** The only source of aluminum-26 is massive stars that explode at the end of their lives.
    1. In addition to creating aluminum-26, such explosions also produce massive shock waves that might trigger the collapse of the Sun's natal cloud.
    2. Also, such explosions often leave behind the collapsed core of the now-dead star, called a neutron star, which can produce the circularly polarized light we speculated earlier is the source of the left-handed molecules of life.

Suggested Reading:

McBride and Gilmour, *An Introduction to the Solar System.*

Woolfson, *The Formation of the Solar System.*

Questions to Consider:

1. Is a nearby exploding star necessarily a rare event in a region of star formation?
2. Without a nearby exploding star, what would have been different in our solar system?

Lecture Seventeen—Transcript
What Happened before the Sun Was Born?

We succeeded, in the last lecture, in determining the age of the asteroids to high precision. But where do they fit into the story? And what preceded their formation? Some relatively short-lived radioactive isotopes tell us much about the solar system's early days, provide a hint of what might have made it come into being, and may even solve the mystery of why all life's amino acids are left-handed.

In particular, traces of extinct radioisotopes, long-vanished through decay, provide clues as to how long it took for solid material to condense out of the protosolar nebula. Solar systems form as a spinning cloud of interstellar dust and gas collapses to a central sphere, which makes the star, plus a thin disk of orbiting material left over from that collapse, which forms the planets. We see this process at work today in regions where stars are forming now in our galaxy. This series of spectacular pictures from the Hubble Space Telescope reveals what's going on. I must emphasize, these are pictures of the sky. They're what it actually looks like up there. They're real visible-light pictures. The colors are real. The pictures are formed by taking images in several different filters, with different-colored filters over the lens of the camera, if you will, and then piecing them together in the computer, but they do represent what space actually looks like where clouds of gas are coming together to form stars.

Several things are prominent in all of these features. There is light and there are dark regions. The dark regions are not an absence of stars or an absence of light. The dark regions are where the dust grains—little tiny particles of dust, sooty kind of material that I spoke about earlier—come together in such profusion that they block the light from the background. You can see that clearly in these pictures.

They're often in long strings and filaments in turbulent areas. Recall that this gas began as gas spread throughout space at a million degrees and gradually cooled off. When this gas falls together under its mutual gravitational attraction, it's not surprising that it's a little turbulent and disordered.

We often see in these pictures very thin filaments of this dark, dusty material; those are the solar systems in formation. The pillars of dark gas that we see often may contain dozens or even hundreds of stars, because they weigh thousands of solar masses each in molecular gas and cold dust. But as individual regions start to collapse, because of the random turbulent motion, this piece over here might be rotating in this direction. As something that's rotating collapses, it spins faster. You're all familiar with the figure skater—she flings out her arms at the end of her routine and then goes into the death spiral, not because it looks pretty, but because as you draw your arms in, as you shrink your radius, you spin faster.

That's exactly what's happening in these interstellar clouds. As the material falls, under gravity, towards the big mass in the center, it's rotating in some random direction, that spin just gets faster and faster and faster. And so while the central object, because of its large gravity, becomes a sphere, the material left over becomes a thin disk, a very thin disk that stretches out from that sphere to billions of miles.

When we peer deeper into these regions of star formation, we can actually see these disks being formed. These are pictures from the Orion Nebula, the nearest region of massive star formation going on today, about 1500 light-years, or about 6000 trillion miles from Earth. We see here 4 examples of what we believe are solar systems in formation. In the center of each is a little bright spot, brighter in some, dimmer in the others, always quite red.

Why is it red? For the same reason that the Sun is red as it sets in the west. As the light from the Sun passes through a greater and greater amount of atmosphere, the rays are scattered away and, so rather than coming directly to you, illuminate the whole sky. The scattered rays are most effectively deflected if they're short wavelengths, or blue light, which is why the sky is blue most of the time, but as the atmospheric path gets longer and longer, as the amount of dust and air molecules the light must pass through gets greater and greater, longer and longer wavelengths get scattered, as well. First the green, then the yellow, then the orange, until the only thing coming directly at you from the Sun is red.

That's exactly what's going on here. The light from that central star is surrounded by this dusty disk, and as the light comes out towards you, it's scattered through more material, so the blue light goes off in other directions, and the green light, and the yellow light, and only the red light gets through. The disk itself is dark, black—again, not because there's nothing there, but because there is something there blocking the light from behind. All these little dust particles are packed so densely together that they're blocking all the light from behind.

They are glowing themselves but not in the part of the spectrum our eyes can see. Recall that our eyes are carefully tuned to the part of the spectrum the Sun puts out, with a temperature of 5000°. These dust particles are much colder, maybe a few hundred degrees above absolute zero, and so while they're radiating energy, it's in the infrared part of the spectrum that our eyes cannot detect.

This was the circumstantial evidence that we had, until recently, that other stars may contain planets. But as I mentioned last time, we now have much more direct evidence that solar systems are common. Indeed, many young stars, after they're formed, are seen to still have this thin disk of material around them. Here are 2 examples, 1 seen face-on that's rotating this way, 1 seen edge-on that's rotating this way, of 2 stars, where the central star has been blocked out by a little mask put in front of the telescope, called a coronagraph. All you see left is light from the star that's scattered off the dust particles in the disk. In both instances, you see the size of Neptune's orbit, the most distant planet from the Sun, drawn to scale in the figure. And you can see these dust disks are a couple times the size of Neptune's orbit, just the size to make new solar systems.

Very recently, at the end of 2008, the Hubble Space Telescope took what is the first actual picture of a planet around another star. The star is Fomalhaut, one of the bright stars you can see with the naked eye in the sky. And like many young stars, it's surrounded by a disk of dust. In this case, again, the coronagraph has blotted out the light from the star so you can see the light from the disk.

The astronomers involved in this project took these very deep pictures with the Hubble Space Telescope. On the left, you see a tiny little dot inside the circle that comes from the square in the upper right-hand part of the picture, which might be a background star, might be a slight enhancement of the dust; hard to tell what it is. But when they came back and took the same picture 2 years later, as you can see in the right-hand picture, the little spot has moved. And, indeed, since we can calculate the distance from the parent star and we know the mass of the parent star, we know exactly how far it should have moved in that 2-year period. In other words, how long its year would be if it orbited around the star. And, indeed, it moved precisely the right amount in precisely the right direction. This is a planet orbiting another nearby star, a young star whose dust disk is still mostly there and it's probably still in the process of forming planets, as our solar system formed 4 billion years ago.

Different materials condense out of this nebula of gas and dust and become solids at different temperatures and, thus, at different distances from the parent star. As the star contracts and starts to shine, by processes we'll describe next time, it starts to illuminate this disk and warm it up, but clearly, the closer you are to the star, the warmer it's going to be, and the father away you are, the colder it's going to be. What we'd like to know is when, exactly, in each of these regions of the disk, the dust particles could stick together, first to form little microscopic particles, then to form macroscopic particles, then marbles, then rocks, then boulders, then asteroids, and ultimately come together, coalescing under their mutual gravity to form planets.

We know the basic physics of how these condensation processes occurs for different kinds of elements. But to calculate from first principles exactly when this occurred in detail is essentially not possible. We use, therefore, the radioactive isotope plutonium-244 to provide a direct measure of the time it took this process to proceed. Those of you paying close attention might be a little surprised that I'm using the element plutonium, because it's number 94 in the periodic table. And you're always told, and indeed I have even said, that the naturally occurring elements run from 1 to 92. That's a slight misstatement. Indeed, all the elements in the current periodic table, which run up well past 100, are naturally occurring in that they can be produced in natural processes, which we'll describe in the next couple of lectures. But all of the heavy ones are radioactive. They're produced in tiny amounts and they decay away, so very few of them are found.

Plutonium-244, however, is produced in stars at the end of their lives and has a half-life of 83 million years, which is short compared to the age of the solar system but not 0. This is the most stable form of plutonium. The kind that we use in nuclear reactors today, plutonium-238, has a much shorter half-life of only 88 years. It's used both in nuclear reactors on the ground and to power spacecraft that travel far from the Sun. Spacecraft that orbit the Earth, of course, gain their energy from solar panels, from panels that convert sunlight into electricity. And the Earth is close enough to the Sun that a moderate-sized panel covering the spacecraft can provide all the power that it needs.

But if one wishes to explore the outer solar system, as we have with satellites over the past 30 years, one gets farther and farther from the Sun, and as a consequence, the amount of solar energy gets weaker and weaker, making the solar panel requirements larger and larger to provide enough electricity to the spacecraft. For very large and complicated spacecraft, which want to send back lots of data in the form of beautiful images and atmospheric measurements, like the spacecraft Cassini, which is currently exploring the Saturnian system and had its little capsule parachute onto the moon of Saturn Titan, it requires so much energy and is now so far from the Sun that a nuclear reactor was required. So a very small amount of plutonium-238 was included in the reactor on that spacecraft to give it power. Since the half-life is 88 years, that spacecraft can be powered for a very long time by this small amount of plutonium.

But it's not the plutonium-238, the kind we make and use in reactors, but plutonium-244 that gives us a clue about the early solar system. It decays through 3 alpha decays to thorium-232. Now, thorium-232 occasionally decays itself through a very rare process of spontaneous fission. We've yet to talk about this kind of nuclear transformation. So far, we've focused only on radioactive decay, which involves the spitting out of an electron or a positron, an alpha particle for particles of the nucleus, or a gamma ray. But there's another kind of nuclear reaction that can occur when a nucleus is so unstable that rather than just spitting off 1 or 2 particles to reach a more stable state, it actually splits roughly in half into 2 nuclei which weigh half as much as the original one.

This process is sufficiently violent, but it spits out lots of extra neutrons, as well. Sometimes 3, 4, 5, 6, 7 neutrons go off independently from the 2 chunks of matter that are left behind. The half-life for this spontaneous fission in thorium-232 is 13.9 billion years, almost precisely the age of the universe, meaning there's still lots of it around.

This violent splitting of the nucleus produces 2 important effects. Since hundreds of millions of electron volts are involved—remember, the light our eyes see is about 1 electron volt and results from atoms having electrons jump between energy levels. The nucleus is much more energetic because it's much more tightly packed, and its processes typically involve millions of electron volts. When you involve hundreds of nuclei all participating in some process, you get hundreds of millions of electron volts given off. This event leaves a microscopic track in the meteoritic material from the atom, the 2 halves of which go rocketing off at close to the speed of light. They smash apart the crystal, leaving fractures on a microscopic level that in principle one could measure and learn about the process and when it occurred. This fission track technique is used in dating rocks on the surface of the Earth all the time and even sometimes in dating fossils. However, for things as old as meteorites, these tracks gradually heal themselves as the displaced atoms find their way back to where they belong in the crystal lattice, and so they've faded out almost completely by this time in the solar system's history.

However, the expelled neutrons—often several neutrons coming from each fission—combine with other nuclei to make exotic isotopes, and it's these we can use to understand the original condensation of the solar system. In particular, both iodine-128 and iodine-129, very rare isotopes of iodine, are produced by neutron capture, the neutron combining with the dominant stable isotope of iodine-127. Both heavy isotopes of iodine decay via beta decay to the corresponding isotopes of xenon, 1 step up in the periodic table, xenon-128 and xenon-129.

As a noble gas, xenon never combines with anything to make compounds and should not be present at all in condensed rocky material which was initially molten and, therefore, from which gas would easily escape. When examining the meteorites, however, one finds both xenon-129 and xenon-128 together with iodine-127, from which its now-deceased parents were made. The ratio of xenon-129 to iodine-127, roughly 1 part in 10,000, provides a clock to measure how fast the material condensed. A high ratio implies early formation, because all the xenon-129 that decayed was trapped and locked away as soon as the material condensed. Whereas a low ratio implied that it took a while for the material to come together and trap the xenon; most of the xenon had escaped.

We collect meteorites that fall to Earth and examine their isotopic composition. There are several types of meteorites that are quite distinct from each other because of their distinct formation histories. Iron meteorites are made mostly of metals—and since iron is the most abundant of the metals, mostly of iron—and clearly were melted and then condensed into large globules. The largest of these meteorites we have are about the size of me. There are stony iron meteorites that consist of a mix of iron metal with a composition of silicon, and aluminum, and other kinds of materials that condense into crystals or rock-like materials. These formed in a different location under different conditions.

Finally, there are the stony meteorites that contain almost no metals at all but just consist of rock-like material. Of this latter category, there are kinds that contain just pure rocks and then there are kinds that contain things called chondrules, little tiny things from microscopic up to the size almost of a marble that are spherical globules of silicates which are, by our models at least, the very first things to condense out of our solar nebula: tiny little microscopic to ball-sized things that contain silicate material, silicon and oxygen, the first abundant elements that form together to make solids.

The carbonaceous chondrites are yet a subcategory of these chondritic materials, which is a subclass that contains less than 5% of all meteorites but includes a high degree of both water and organic compounds inside these little globules of silicate material. This implies—since the water is volatile and if it were hot would evaporate, and the organic materials are very delicate compounds—that these globules have never been heated after they were formed. They are, therefore, we conclude, the most primitive and earliest type of meteorite.

They show, when we examine their xenon-to-iodine ratios, a very narrow spread in condensation times all clustered around a value of about 12 million years. And so this is the period that it took for the early solar nebula to begin to condense into solid particles. This provides us with the first quantitative clue about the earliest events in the formation of the planets.

Extinct radioisotopes also provide a clue as to what happened just before the solar system formed. Many iron meteorites show obvious signs of differentiation. You recall, when I discussed the Earth, that the light rocks, 2 to 3 grams per cubic centimeter, exist on the surface and the heavy metals have all sunk to the core. When the Earth was molten, like oil and vinegar, the light ones rose to the top and the heavy ones sank to the bottom. In these meteorites, which are small—you know, this kind of size—the same thing occurred. The heavy metals are on the inside and the light minerals are on the surface.

This provides strong evidence that after they came together, after they solidified, long after the first 12 million years when the chondrules were formed, something melted the asteroids. Now, on Earth, the heating that leads to the Earth being molten and the fluidity which allows these atoms of different mass to separate from each other is just a consequence of gravity. Things falling on to the Earth, like the asteroid that killed the dinosaurs, come in with tremendous speed and produce an enormous amount of heat just from converting gravitational energy to heat energy, the jiggling of atoms. But asteroids, even the ones a couple hundred kilometers across, let alone the ones that are the size that would fit in this room, have way too little gravitational energy to have melted the material. That can't be the explanation for how they differentiated with the heavy metals on the inside and the light materials on the outside.

An alternative is radioactive decay. This is also a heating source in that when a nucleus undergoes a transformation, it spits out a very high-energy particle, typically with energies of millions of electron volts. That particle runs along, rattling off atoms in the material in which it was formed, and as it bounces off, it transfers some of its energy. What does that mean? It means those molecules that it hit jiggle faster. What does that mean? That means it's hotter. Indeed, if enough energy is released, it's so hot that the rigid bonds holding a solid in place are broken and it becomes a liquid. In a liquid, the heavy molecules can sink to the center and the light molecules can rise to the surface.

If radioactive decay is to be the explanation for differentiation in these asteroids, we need it to be an abundant species and we need it to have the right kind of half-life to have decayed away over the early part of the solar system's history. Aluminum, it turns out, is one of the most abundant elements that is produced in stars, outside of hydrogen and helium, making up over 2% of all the heavier elements.

Aluminum-26 is a radioactive isotope which, when it's produced in the cores of stars, constitutes about 50 atoms out of every million aluminum atoms produced. Aluminum-27 is the stable one, which completely dominates the amount of aluminum made. Aluminum-26, however, undergoes beta decay with a half-life of only 730,000 years, releasing 4 million electron volts for every decay.

Now, we can do a simple little calculation. Aluminum-26 is 50 out of every million atoms of aluminum. And aluminum is 2% of all the elements, again, leaving aside hydrogen and helium, which don't participate in the formation of asteroids, helium because it's a noble a gas, hydrogen because it doesn't condense until very low temperatures. And so, 2%, 0.02×50 atoms out of every million means that 1 out of every million atoms that condenses into a solid form will be aluminum-26.

It takes considerably less than 4 electron volts to break the bonds that hold the solid material in the asteroid in place—in other words, to melt that material—so the solid, where the atoms are rigidly locked, turns into a liquid, where they can flow over each other and the material can differentiate. Four electron volts per atom, and yet we have 1 in a million atoms being aluminum, but each aluminum decay produces 4 million electron volts of energy. And so we have just enough energy in the aluminum-26 alone to melt the asteroids and explain their mysterious differentiation.

But, in fact, it's not just this indirect evidence of differentiated asteroids. There's more solid evidence that this short-lived radioactive isotope was indeed present at the birth of the solar system. Aluminum-26—aluminum being number 13 in the periodic table—undergoes a decay to magnesium-26, number 12 in the periodic table, by spitting off a positron and the inevitable neutrino. Now, it turns out, 11% of all the magnesium on Earth is magnesium-26. But in the chondrules, what we do is use a technique similar to that I discussed last time of making a ratio from magnesium-26 to the stable isotope magnesium-24 and aluminum-27, the stable isotope of aluminum, to magnesium-24. Magnesium-26, remember, is the product of the aluminum-26 decay. It turns out there's a beautiful correlation between the amount of magnesium-26 in excess of what one would expect with the amount of aluminum-27 that's present in these chondrules, confirming that magnesium-26 is indeed from the decay of the sister isotope, aluminum-26. Where there is more aluminum, there is more magnesium-26; where there is less aluminum, there is less magnesium-26, just as we'd expect if the excess magnesium-26 all came from this decay.

The question arises: Where did this short-lived radioactive isotope come from? It must have been produced within less than 10 half-lives or so, or less than 10 million years of the solar system's birth.

Because after 10 half-lives, $\frac{1}{2} \times \frac{1}{2} \times \frac{1}{2} \times \frac{1}{2} \times \frac{1}{2} \times \frac{1}{2} \ldots$, means that we're down to only $\frac{1}{1000}$ of the original abundance. And that wouldn't be nearly enough to have melted the asteroids. So this radioactive isotope had to appear on the scene at the time the solar system came into being—right at the same time, not eons earlier, produced in some distant other star.

The primary source of aluminum-26 is massive stars which explode at the end of their lives. As we'll learn in the next 2 lectures, stars are the factories which produce atoms. All of the atoms, besides hydrogen and helium, were cooked up inside stars, and this applies to the stable isotopes, as well as to the radioactive ones, like aluminum-26.

In addition to creating things like aluminum-26, these massive stars end their lives with huge explosions, which disgorge all the atoms they've cooked up over their lifetime, but also lead to a blast wave that rockets through the interstellar medium—a sonic boom, if you will—crashing into any clouds in the way and quite possibly triggering those clouds to start to collapse. The pressure wave pushes matter together. Once there's an excess concentration of matter, gravity can take over and material can fall from all different directions, building it up to the size of a star. But this trigger of a stellar explosion might be essential to producing stars. Indeed, when we look in the Orion Nebula today, this nearby region of star formation, we see directly the remnants of exploded stars running into those clouds of material that have yet formed stars and compressing them in just such a way to trigger the collapse of star formation. Indeed, looking in distant galaxies, we can see the overall architecture of the galaxy is governed by the explosions of these stars.

In addition to producing radioactive isotopes like aluminum-26 and in producing the shock wave that might have triggered the collapse of the solar system, such an explosion also often leaves behind the collapsed core of the now-dead star in the form of one of these bizarre objects called a neutron star. The core of the star collapses so rapidly that the electrons are driven into the nuclei of the atom, combining with protons. An electron plus a proton makes a neutron, the other neutrons are still there, and the entire body of the star is principally neutrons.

The collapse, as I said earlier, leads anything that's spinning to spin faster, and so these objects spin very rapidly. In addition, the collapse amplifies the magnetic field that would have been present in the star, leading to the extremely rapidly rotating, magnetized, incredibly dense object that could have illuminated the early solar system with the circularly polarized light that preferentially destroyed the right-handed amino acids, the molecules with right-hand symmetry, leaving an excess of left-handed molecules, which came to yield life today on our planet.

Reconstructing the origins of our geology and our biology at a distance of 4.56 billion years is a fascinating task. But the formation of the Sun and the Earth is far from the beginning—indeed, over $\frac{2}{3}$ of the life of the universe had already passed when this occurred. What came earlier? Indeed, a more fundamental question: Where did the atoms in the pre-solar cloud come from? To that answer question, we must turn to examine the lives of the star, a story that we'll undertake next time.

Lecture Eighteen
Atoms Are Star Stuff—Cooking Up Carbon

Scope:

When the universe began, it contained only 2 kinds of atoms: hydrogen and helium. The other 90 atom types (plus some of the helium) in the universe today have been cooked up inside stars. We must begin by asking: What is a star? The astrophysicist's retort—a gravitationally bound ball of plasma supported by thermal pressure in hydrostatic equilibrium, emitting blackbody radiation generated by nuclear fusion—is quite simple when explained and allows us to understand how stars of different masses generate different elements and distribute them to interstellar space.

Outline

I. To an astrophysicist, a star is "a plasma, gravitationally bound, supported by thermal pressure in hydrostatic equilibrium (usually), emitting blackbody radiation, and powered by nuclear fusion."

  A. Recall from Lecture Two that a plasma is the fourth state of matter, in which electrons are largely detached from their nuclei.
   1. Throughout most of their mass, stars are pure plasmas consisting only of bare nuclei and free electrons.
   2. In the outermost atmosphere, the temperature may be low enough, and the collisions nonviolent enough, to allow some the electrons to recombine and make (usually ionized) atoms.
   3. These atoms of the outer layers reveal the star's chemical composition, rotation rate, atmospheric structure, temperature, magnetic field strength, and surface gravity.
   4. The presence or absence of radioactive atoms can be a measure of a star's age, although this is not the primary means for dating stars.

  B. Recall from Lecture Three that gravity is 1 of the 4 fundamental forces of nature, which here determines a star's structure.
   1. Although gravity is the weakest force, the mass of a star is so large that gravity dominates its structure.

2. Stellar masses range from 0.08 to roughly 80 times the mass of the Sun, where 1 solar mass = 4000 trillion trillion tons, or 300,000 Earths.
3. Since the strength of gravity depends only on distance between the attracting particles, stars are spherical (with slight distortions from rotation and/or tides raised by companions).

C. Also, recall from Lecture Three that "heat" is simply the energy of the motion of individual particles; thermal pressure is the force exerted by those particles as they collide with each other.
1. The average speed of a particle is given by a simple relation between particle mass and temperature.
2. When tires get hot in the summer, the pressure rises—hot means the air molecules move faster, thus colliding with the tire walls more energetically and more frequently, pushing them outward.
3. Subatomic particles in a stellar core similarly push outward on the material above.
4. The surface temperatures of stars range from 3000 to 50,000 on the Kelvin scale.
5. The central temperatures range from 15,000,000 K to more than 1,000,000,000 K.

D. Hydrostatic equilibrium represents the balance between the 2 forces acting in a star: gravity pulling ever inward, and thermal pressure pushing ever outward.
1. The high central temperatures help support the huge burden of overlying mass pushing in.
2. Stellar models are constructed by starting with the temperature and pressure in the outermost layer (which we can observe directly) and calculating what temperature and pressure the next layer down must have to support it, and so forth.
3. Throughout more than 90% of their lives, stars maintain the equilibrium. When things get out of balance, the star must either expand or contract.

E. We see stars because they radiate light energy into space.
1. Recall from Lecture Five that moving charges radiate electromagnetic radiation.
2. The wavelength (or color) of radiation given off is related to the energy of the emitting particles.

3. The Sun has a surface temperature of 5800 K, meaning that its radiation maximum is at the wavelength of yellow light.
4. Cooler stars are redder, hotter stars bluer.
5. The range of energies of the radiating particles leads to a range of wavelengths of light, the distribution of which we call a blackbody spectrum.
6. The Sun radiates at the rate of 400 trillion trillion watts.
7. The energy leaving a star would cool it unless the temperature were maintained by some other process.

F. Nuclear fusion is the source of energy that keeps stars shining and in hydrostatic equilibrium.

II. Fusion is the process by which atomic nuclei stick together to make new nuclei.
   A. Recall that the strong force overcomes electrostatic repulsion at small distances.
      1. If protons and neutrons can be brought close enough to each other, they snap together.
      2. This process gives off energy and creates a new isotope or element.
      3. Since the nuclear force is so powerful, the energy release is huge—1 to 10 million times that of the electromagnetic interactions involved in chemical reactions.
      4. The typical nuclear reaction emits several mega–electron volts per nucleon involved.
   B. Recall also from Lecture Five that nuclear stability depends on the ratio of neutrons to protons in a nucleus.
      1. Thus some combinations are more stable than others.
      2. The more stable they are, the more tightly they are bound, and the more energy they give off in formation.
      3. The most stable arrangement possible is of 26 protons and 30 neutrons, the iron nucleus.
      4. The nuclear binding energy curve represents the stability of the nuclei and allows one to calculate the energy released in any reaction.
      5. Combining 4 hydrogen nuclei to make helium in a 3-step process is the reaction that powers most stars for the majority of their lives.

III. To make heavier elements, the electric repulsion to overcome is greater.
   A. The repulsion is proportional to the product of the charges on each nucleus.
      1. To stick 2 helium nuclei together is 4 times harder than joining 2 hydrogen nuclei.
      2. Furthermore, boron-8 (2 helium nuclei) is highly unstable and falls apart in less than 10^{-16} seconds, creating a bottleneck in the fusion process.
   B. The triple-alpha (3-helium) reaction brings 3 nuclei together to make the highly stable carbon nucleus.
      1. This reaction is the source of most of the carbon in the universe.
      2. This is the reaction that powers stars after hydrogen exhaustion.
      3. This reaction requires much higher temperatures and thus requires a readjustment of the star's equilibrium.
   C. The need for a succession of new fuel sources and the readjustments they require provide the story of stellar evolution.

Suggested Reading:
Bennett, Donahue, Schneider, and Voit, *The Essential Cosmic Perspective*.

Questions to Consider:
1. What are some of the desiderata of interest in thinking about which types of stars might support planets with life on them?
2. How is the Milky Way changing over cosmic times?

Lecture Eighteen—Transcript
Atoms Are Star Stuff—Cooking Up Carbon

The question we ended with last time, where did the atoms that formed the solar system come from, has 2 very distinct answers. Virtually all of the hydrogen and 80% of the helium was formed in the first few minutes of the universe, a time we will explore in a future lecture. All the rest, however—the carbon, nitrogen, and oxygen of life; the gold, silver, and platinum; the neon; the iron; and the uranium—were all were made by stars. It is important, then, that we turn our attention to the processes which give this essential richness to the atomic world, and we begin by asking the simple question: What is a star?

I must confess this question would not have occurred to me unbidden. I've lived with the knowledge of what stars are for so long that it seems to me as obvious as asking the question: What is water? Or what is a finger? You know what those are because you've known them all your life. The importance of this question, however, for most people, was brought home to me by an incident that occurred a few years ago.

I grew up in a small town in southeastern Massachusetts, which I considered a very boring place and ran screaming from it at the age of 17, rarely to visit it again. However, a few years ago, I went to visit my late-80-something father, who by that point had joined an organization consisting mostly of other 80-something men in his town, which was in need of entertainment. And so he asked if I would come and give a lecture on what I did. To this event, he invited my two 80-something maiden aunts, who sat quietly in the back of the room and listened.

I was giving a talk on neutron stars, a subject that I'm fascinated with and have spent most of my career studying, and in the course of this talk, I said something to the effect that "Well, a star is a ball of gas," and then went on to describe the star. And Sophie and Edith, my 2 aunts, came up to me, rather wide-eyed at the end of my lecture, and said, "Is that what a star is? A ball of gas?"

It, again, seemed obvious to me that everyone should know what a star was just because I knew what a star was. But, of course, it's not obvious at all. What you see as a star is a tiny twinkling spot of light that appears night after night in the sky. It's not obvious what that is, and it's certainly not obvious that it's a ball of gas. Furthermore, my statement was slightly incorrect, as you'll see.

To an astrophysicist, a star is, and I quote, "a plasma, gravitationally bound, supported by thermal pressure in hydrostatic equilibrium, emitting blackbody radiation and powered by nuclear fusion." Obviously, this definition was going to be of little help to Sophie and Edith and probably isn't much help to you, too. Although, over the course of the lectures, many of those terms have been introduced.

Recall for example, from Lecture Two, that a plasma is the fourth state of matter, in which electrons are largely detached from their nuclei. So atoms, as such, don't exist, but the constituents of atoms, the electrons and the nuclei, are running around independently of each other. Throughout most of its mass, stars are pure plasmas consisting only of bare nuclei and free electrons.

In the outermost atmosphere of the star, the temperature may be low enough—that is, a mere 5000° or 10,000°—that the collisions are not quite violent enough to always split an electron that happens to join up with a positive nucleus apart. Some electrons recombine and make what are typically ionized atoms—that is, atoms lacking some of their electrons—but not all of them. With electrons in their orbits around a nucleus, they can make transitions to lower and higher orbits and, in the process, give off and absorb light.

It is these signatures—again, unique to each atom—that reveal the chemical composition of the star (something we would otherwise not know), the rotation rate of the star from the broadening of the spectral lines, the atmospheric structure of the star, the temperature of the star, the magnetic field strength of the star, and the surface gravity of the star, which gives us a clue as to its mass and size. All critical inputs to allow us to construct models of stars. So the presence in the atmosphere of a few ionized atoms is key to our understanding the physics of stars.

The presence or absence of radioactive atoms can sometimes be detected in a star and can give us a measure of age, but this is not the primary means we have for dating stars. "A plasma," then, "gravitationally bound": Recall from Lecture Three that gravity is 1 of the 4 fundamental forces of nature, which here determines a star's structure. Although gravity is, as I said, by far the weakest of the forces, the enormous mass of the star is so great that gravity completely dominates its structure.

Stellar masses range from about 8% the mass of our Sun to around 80 times the mass of our Sun. In other words our Sun is a very typical star, right in the middle of the mass range. One solar mass, the mass of our Sun, is 2 thousand trillion trillion tons, or about 300,000 times the mass of the Earth.

Now, why can't we have stars, if they're just balls of plasma, of any size? Why is the limit at 8% the mass of the Sun and 80 times the mass of the Sun? For 2 different reasons. To power a star, as we'll see, we have to squeeze the core to such a high pressure and density that the nuclei of hydrogen atoms, protons, can stick together and make helium. At 8% the mass of the Sun, a star can just barely generate the pressure and temperature necessary for this reaction to occur. At 7% the mass of the Sun, there's just simply not enough overlying material; the gravitational force is simply too weak to make this process occur. And so since a star is, by definition, one that is burning nuclear fuel, less than 8% of a solar mass isn't, by definition, a star.

The upper end, 80 times the mass of the star, is because too much energy, rather than not enough, is being generated. If we go today to a star of 100 or 200 times the mass of the Sun, the energy generated in its core is so great that it will simply blow the star to pieces, and so a stable star can't exist. Thus, we have this range of a factor of 1000 from 8% the mass of the Sun to 80 times the mass of the Sun that encompasses all of the hundreds of billions of stars in our galaxy.

Since the strength of gravity depends only on the distance between the 2 attracting particles, stars become spherical with some slight distortions if they're rotating very rapidly, where they bulge out at the center, or if they have a companion, which raises tides on the star the way the Moon raises tides on Earth.

So a star is a plasma that's gravitationally bound. The next part of the definition says that it's supported by thermal pressure. What do I mean by that? Also recall from Lecture Three that heat is simply the energy of the motion of individual particles, and thermal pressure is the force exerted by those particles as they collide with each other and try to get away, that is, expand from the center of the star.

The average speed of a particle, you will recall, is proportional to the square root of the temperature divided by the particle's mass; more massive particles move slowly, higher temperatures make particles move faster. When your tires on your car, for example, get hot in the summer, they expand. What does "hot" mean? Hot means the air molecules inside your tire are moving around faster, meaning they both collide with the tire walls more frequently and each collision is more energetic. As a consequence, the pressure of this gas expands the tire.

Likewise, the pressure of all the subatomic particles moving around in the stellar core are pushing outward on the material while gravity tries to bring it in. The surface temperatures of stars range from about 3000°, for those lowest-mass stars only 8% the mass of the Sun to over 50,000° at the upper end. All temperatures, of course, are on this Kelvin or absolute physical scale, where 0 represents no motion. So 3000° to 50,000°—that's on the surface of the stars. But to make particles go fast enough to fuse, as we'll see they need to do, the central temperatures of stars are much higher, ranging from 15 million degrees—that is, in a star such as the Sun—to over a billion degrees Kelvin.

"A plasma, gravitationally bound, supported by thermal pressure"—all that begins to make sense. "In hydrostatic equilibrium"—or, at least, usually so—what do I mean by that? Hydrostatic equilibrium represents the balance between the 2 forces acting in the star. Gravity, which is always trying to crush the star—every particle with mass always attracts other particles with mass and they always want to get closer together, so gravity, which starts out making a star from a cloud of gas, shrinks it down and wants to keep shrinking all the way to the center. But when it gets hot enough inside, when the particles are running around fast enough, the thermal pressure they exert pushes outward. Most stars, more than 90% of stars—and our Sun is an example—are in a dynamic balance between these 2 forces: gravity that's trying to crush the star and thermal pressure which is trying to expand it. And so, every morning when you get up, the Sun in the sky is exactly the same size because these 2 forces are in balance. And they've been in balance for 4.5 billion years, which is why the conditions on Earth have remained roughly constant over that time, allowing life to evolve. The high central temperatures in the star help in supporting this huge burden of overlying gas.

To understand how stars work, we construct models on computers that mean to represent how the star behaves. We do this in a sequential fashion. We start with the outer layer of the star, which we can see directly. We, therefore, know its temperature very precisely. We know the density of atoms there and, therefore, the total mass in the thin shell on the outside of the star, the part we directly observe.

We then calculate what temperature and pressure and, therefore, density of material must exist in the thin layer right under the surface to support that surface. If the pressure is too high, the temperature is too high because the atoms are running around too fast, or the density is too high—there's too many atoms pushing out—then the outer layer of the star will lift off in our computer model. That's no good. If the pressure is too low—that is, there aren't enough particles there; the density's too low—or the temperature is too low so the particles aren't moving fast enough, then the outer layer of the star in our model will collapse. And that's not good. And so we adjust the layer 1 step down from the outside of the star to support the layer above it.

Now we're on our way, because we've got the outer layer and the next layer below it consistent with that. What supports that layer? Of course, the layer below that. And so we take the next layer, calculating the temperature, pressure, and density such that it supports layer 2. And then to layer 4, supporting layer 3, etc., all the way to the core of the star. We have a test of this, which I'll describe at the end, which allows us to confirm this model-building process.

Throughout more than 90% of their lives, stars maintain this delicate equilibrium between pressure pushing out and gravity pushing in. When things get out of balance, however, the star must either expand, if the pressure inside is too great, or contract, if the pressure inside is not great enough. We'll talk about those processes next time, because both are critical in the lives of the stars and the creation of the elements.

So, "a plasma, gravitationally bound, supported by thermal pressure in hydrostatic equilibrium, emitting blackbody radiation"; that's a mouthful. What do we mean there? We see stars because they radiate light energy into space. Recall from Lecture Four—you see how all those early lectures were preparation to what we're building up to here—that moving charges radiate electromagnetic radiation. Positive and negative charges zipping around each other, as they are inside a star and on the surface of a star, give off waves of electromagnetic radiation, the stuff that we see, we call light. The wavelength, or color of the radiation, given off is related to the energy of the emitting particles. The faster the zipping around, the higher the frequency or shorter the wavelength of the radiation and, in the case of our eyes, the bluer the color. The slower they're moving, the longer the wavelength.

We learned, of course, that the wavelengths that we see are a tiny fraction, only 2%, of all the waves possible from radiating charges. The Sun has a surface temperature of 5800°, meaning that the maximum of its radiation is a wavelength in the yellow-green part of the spectrum. Cooler stars, which means their atmospheric molecules are moving more slowly, are redder, and hotter stars bluer. If you're very acute and you're in a very dark place, you can actually see a slight tinge to the color of the stars; some of them are bluer and some of them are redder. The stars at the head and foot of Orion, the constellation one can see in the winter sky, are a good example of this.

The range of energies of the radiating particles leads to a range of wavelengths of the light. As you recall, particles are not moving all exactly the same speed at some temperature; the average speed is given by the square root of temperature over mass, but there's a whole range of speeds. And when we talked about the Earth's atmosphere, we saw that the fastest ones escaped. Likewise, in a star, there are fast particles and slow particles all consistent with a given temperature. And since the fast ones radiate short wavelengths of light and the slow ones radiate long wavelengths of light, we get a distribution of wavelengths, a distribution of colors. And that distribution has a very particular form related directly to the particular distribution of the radiating particles. We call that a blackbody spectrum, a spectrum of different wavelengths distributed in a certain way.

In the case of our Sun, the temperature, 5800°, defines the peak of that spectrum as being in the yellow-green part of the spectrum. It's not an accident that our eyes are most sensitive to yellow and green light. We have, over 400 million years of evolution, evolved a device, the thing we call our eyes, that's tuned very precisely to the wavelength the Sun puts out most of. The Sun puts out a little bit of infrared light and an even tinier amount of radio waves. It also puts out a little bit of ultraviolet light and an even tinier amount of x-rays. But our eyes are simply not sensitive to them because the most efficient thing for our sensors would be to be where most of the radiation is, and that's in the visible light part of the spectrum.

There are stars in the universe that emit most of their energy in the x-ray part of the spectrum. That would have been a sensible place to put the planet on which Superman grew up, for example, because then naturally, Superman would have x-ray vision—again, evolution having tuned his eyes to be sensitive to the bulk of the radiation that his parent star puts out. He wouldn't have been born on an asteroid orbiting our Sun.

The Sun radiates at the rate of 400 trillion trillion watts. That's a lot of light bulbs. The energy leaving the star would cool it off, because energy going away means something gets cooler. The atoms give up energy—they slow down—unless this temperature were maintained somehow by some other process. And now we're nearing the end of our definition: "a plasma, gravitationally bound, supported by thermal pressure in hydrostatic equilibrium, and emitting blackbody radiation, powered by nuclear fusion." It must be powered by something if it's emitting all that energy.

Nuclear fusion is the source of energy that keeps stars shining and maintains their hydrostatic equilibrium. Note it's not chemistry. The connection of atoms to each other, when electrons are shared between them in a molecule, gives off energies measured in electron volts. In a star, first of all, there are no atoms—the electrons and protons are divorced—but the reactions occur between the protons at scales 10,000 times smaller than the electron orbits the atom, with energies millions of times greater, such that each reaction gives off millions of electron volts, not just electron volts. This is Lord Kelvin's "unknown process in the storehouse of creation": nuclear reactions, which provide so much more energy they can keep the Sun shining for many billions of years.

Fusion is a process we haven't encountered yet, by which atomic nuclei actually stick together and make new nuclei. Not a simple transformation, as in radioactive decay, where 1 or 2 particles fly out of the nucleus, but the construction of a new nucleus by the fusing together of the nuclear particles, protons and neutrons. Recall that the strong force overcomes the electrostatic repulsion that 2 positively charged particles would feel once you get to small enough distances. Out here, these 2 positive particles will sense each other and fly apart because electric repulsion is strong. But if you get them close enough, into a distance of 10^{-14} meters, the realm of the strong force, the strong force overwhelms the electromagnetic force and forces the particles to stick together.

If protons and neutrons can be brought close enough, then, they snap together, giving off an enormous amount of energy and creating a new isotope or a new element. Since the nuclear force is so powerful, the energy released is huge, from 1 to 10 million times that of the electron-proton interactions that occur in chemistry. The typical nuclear reaction emits several million electron volts for each nucleon involved. Recall from Lecture Five that nuclear stability depends on the ratio of protons to neutrons in a nucleus. Thus, some combinations are more stable than others. And the more stable they are, the more tightly they are bound, the more energy they give off when they're formed.

It turns out the most stable arrangement possible is the 26 protons and 30 neutrons which make up an iron nucleus, iron-56. Now, again, this is not to be confused with chemical stability as, for example, in the noble gases, which don't combine with anything because their electron orbits are perfectly happy. This is nuclear binding energy, the nuclear particles down on this tiny scale of the nucleus, where the nuclear strong force is the dominant entity and in which we must arrange protons and neutrons in just such a way that that strong force maximizes stability.

Each different isotope of each different element has a particular binding energy. Putting these all on 1 graph gives us what we call the nuclear binding energy curve, the stability of nuclei. On one axis, the y-axis, we plot the binding energy per nucleon—the amount of energy I get if I stick all those protons and neutrons together. Hydrogen, of course, is 0, because hydrogen has 1 proton and nothing to bind to, so clearly there's no binding energy. Helium is pretty far down on the curve, several million electron volts for each helium nucleon bound together in a helium nucleus, 2 protons and 2 neutrons bound together, 4 particles, each of them giving off several million electron volts when they snap into place to make a helium nucleus. And so on, down through the light elements in the periodic table, all the way up to iron, where iron, with 56 nuclear particles, represents the ultimate in a stable nucleus.

As we add more particles, more protons and more neutrons, to make all the other 66 elements beyond iron, we actually are making slightly less stable nuclei. This is one reason why when we get all the way to the end of the process, out near uranium, none of the isotopes is stable. They are all falling apart due to radioactive decay. Nonetheless, in creating these elements, energy is given off, because each time I snap another proton or neutron into place, I get energy. The process that most stars use to shine is the simplest one of all, combining 4 hydrogen nuclei together to make helium. This is a 3-step process and it's the reaction that powers all stars for the majority, but not all, of their lifetimes.

We start off in the center of the Sun with 2 protons—2 hydrogen nuclei; they're simply protons. Now, again, when they try to come together, they feel the repulsion of their 2 positive charges and tend to fly apart again. The center of the Sun is enormously dense, and at a temperature of 15 million degrees, the particles are moving incredibly rapidly. And so there are millions of collisions per second between one proton and another. But in almost none of these, that is, only once in 10 billion years on average, with 10 million collisions per second going on—so it's enormously improbable—do these particles actually get close enough that they can snap together and stick.

Now, once in 10 billion years—you'd think it would never happen. But there are 10^{57} protons inside the Sun. So while each one has a really low probability of this happening, altogether it happens a lot—every second. 2 protons coming together to make—what? Proton plus proton would make a positive charge of 2, and would that be stable? No, we make deuterium. In the process, in the violent collision and the snapping together through the nuclear force, one of the protons is turned into a neutron and we get deuterium, the heavy form of hydrogen, a nucleus with one proton and one neutron. That, of course, means we have to balance the charge. We started out with 2 positive charges; we end up with 1. What happened to the other one? The other one goes off in a classic beta decay with an antielectron, a positively charged electron, flying away along with its ever-present accompanying neutrino. So hydrogen plus hydrogen come together to make deuterium, plus a positron, which goes off and annihilates with some electron somewhere, and a neutrino, which streams directly out of the Sun because it interacts so poorly with other particles.

Now we've got a deuterium nucleus; what happens to it? Deuterium, it turns out, is highly reactive. As soon as it finds another proton—which will be almost instantaneously because the collisions are so frequent—it gobbles that up and they snap together. Now what have we got? A proton plus a neutron in the deuterium nucleus plus another proton added means 2 positive charges, 3 nuclear particles: We get the light form of helium, helium-3, so helium, with 2 protons and 1 neutron, an atomic weight of 3. And in that process, again, the snapping together gives off energy, so gamma rays go flying off from the Sun, the ultimate source of the light and heat we get from its surface.

Now we've got a helium-3 nucleus. What happens to it? Well, it wanders around in the Sun until it finally finds another helium-3 nucleus which sticks together. Now we've got 6 particles, but what are we left with? That turns out to be unstable; 2 of the protons go flying away, and we end up with the ultimate product of this fusion reaction, helium-4. Two protons and 2 neutrons—an alpha particle—with 2 other protons reinjected back into the Sun to join the game again. Plus, as always when nuclear particles stick, we get more energy pouring out of the Sun.

Why should we believe this picture? I'm telling you about a place that's 15 million degrees, with particles colliding millions of times a second, only sticking once every 10 billion years, yet nonetheless coming together and producing a new kind of element, helium, out of hydrogen in a 3-step process. We can't possibly see what's going on in the center of the star—we can't see with it visible light; we can't even see the gamma rays that are produced—but we can see what's going on with one of the byproducts of these processes. And that is the neutrinos.

Neutrinos, you'll recall, only interact through the weak nuclear interaction, and that interaction is weak indeed. Neutrinos can pass right through the Sun, just as though it weren't there at all, through the intervening hundred million miles of space between us and the Sun, and be captured at Earth. But it's the capturing them that's the problem, because neutrinos pass right through the Earth, as well. Nonetheless, undaunted by this difficulty, in the 1960s, some scientists built a detector in order to try to catch solar neutrinos to see what was going on in the center of the Sun.

It turns out it's hard. With a tank containing several hundred thousand gallons of cleaning fluid, carbon tetrachloride, once every month or so, a few of the chlorine atoms would get turned into argon atoms, argon nuclei that were, in fact, radioactive and could be detected through their radioactive decay. Year after year, decade after decade, Ray Davis filled his tank of chlorine deep in a gold mine in the Dakotas and, once a month, cleaned it out, looking for this tiny number of argon atoms transformed by solar neutrinos. Remarkably, he found them, but he didn't find enough. He only found a third as many neutrinos as he expected to see. This caused great consternation in the astronomical community, because we thought we'd figured out the stars. We thought we could predict to a few percent how many neutrinos should be coming out, simply by observing how much energy was coming off the surface and knowing how much energy each reaction in its core must produce.

And yet we were off by a factor of 3. This was an astounding result and, for decades, caused a vigorous debate between astronomers and physicists. The astronomers, on the one hand, convinced their models were right, said there must something wrong with the physics of neutrinos. The physicists, with their typical arrogance, said we understand neutrinos; there's clearly something wrong with the technical details of your messy models that attempt to understand stars.

I wouldn't be telling you this story as an astronomer if we hadn't won in the end. It turned out, less than a decade ago, that the problem is with the neutrinos. As you'll recall from the beginning lectures, there are 3 different kinds of neutrinos, corresponding to each of the different generations of leptons or light particles: the electron neutrino, the muon neutrino, and the tau neutrino. It turns out, contrary to the expectations of the standard model, that neutrinos can spontaneously morph from one of these kinds to another: An electron neutrino can become a muon neutrino, a muon neutrino can become a tau neutrino, and vice versa. What the problem was in Davis's experiment is that the neutrinos on their way from the Sun were spontaneously transforming themselves from the type of electron neutrinos that his detector was designed to detect into the other 2 types, muon neutrinos and tau neutrinos. And so the factor of 3 that we were off was precisely right. Because a third of them were the kind he was looking for, and the $\frac{2}{3}$ were the other 2 kinds. Just perfect. This has now been confirmed by observations that can detect the other 2 kinds of neutrinos, and thus, we are quite confident that the models we have built to understand the stars and the elements they create are correct.

To make heavier elements, it's even harder, because the electric repulsion you have to overcome becomes greater. The repulsion is proportional to the product of the charges on each of the 2 nuclei. So to stick 2 helium nuclei together with 2 plus charges here and 2 plus charges here is 4 times harder than joining 2 hydrogen nuclei together. Furthermore, joining 2 helium-4 nuclei together, in the violence of the collision, produces boron-8, an isotope which has 5 protons and only 3 neutrons and, therefore, is off the nuclear stability line and, therefore, unstable. In fact, it's so unstable it falls apart in less than 10^{-16} seconds, which creates a bottleneck in the fusion process, because we can go from hydrogen to deuterium, deuterium to light helium, light helium to heavy helium, but when we try to make the next step and fuse helium together, we're stuck.

What happens is that the core of the star readjusts such that the helium is in such close proximity that 3 heliums can suddenly stick together, the so-called triple-alpha reaction which brings 3 heliums together to make carbon. Two protons and 2 neutrons plus 2 protons and 2 neutrons plus 2 protons and 2 neutrons makes 6 protons and 6 neutrons, or the highly stable nucleus of carbon. This reaction is the source of most of the carbon in the universe. And this is the reaction that powers the stars after they exhaust their hydrogen fuel. It requires much higher temperatures and densities, however, and this requires a readjustment in the star's equilibrium.

In the case of our Sun, this readjustment will be fatal for life on Earth, because shortly thereafter, the Sun will swell up and the Earth will be orbiting through its atmosphere, a most unpleasant experience. Shortly after that, the Sun will run out of its helium fuel and it will also die. More massive stars, however, have an interesting story. They go through multiple cycles of element production, making ever-more oxygen, and carbon, and magnesium, and sulfur, and iron, and all the other elements necessary for life.

Now you know what a star is: a plasma, gravitationally bound, supported by thermal pressure in hydrostatic equilibrium, usually emitting blackbody radiation and powered by nuclear fusion.

Does it make them any less romantic to know how they work? I, for one, don't think so. They are not just pinpricks in a velvet cloth letting the supernal light shine through. They're element factories, taking boring hydrogen and helium nuclei and constructing the elements of which our variegated world is made.

To complete our atomic history, however, we must first make sure that stars produce the various elements that we find in their observed relative abundances: oxygen, there's lots of; platinum is rare. And then, we must arrange for a method by which these newly created elements are distributed to interstellar space, ready to be incorporated into new generations of stars and planets. These are the stories we weave together next time in recounting what we know of the lives of the stars.

Lecture Nineteen
The Lives of Big Stars—Cooking Up Big Atoms

Scope:

While the placid, unchanging night sky gives the impression of an infinite and immutable universe, this impression is profoundly wrong. In fact, stars are born, live out their lives, pass through a rather long and boring middle age, and then die in a life cycle that is, in fact, better understood than the human life cycle. The time scales for stellar evolution are measured in millions to billions of years, much longer than human life expectancy. Nonetheless, by assaying the properties of hundreds of thousands of stars, we can reconstruct their life cycles in remarkable detail and infer quantitatively how they produced all of the elements, from carbon to uranium, that we find in the galaxy today.

Outline

I. Each star we can see seems to maintain a constant position, brightness, and color from night to night, from century to century.
   - **A.** Stars are, in fact, moving through the sky, but their enormous distances (the closest is 26 trillion miles away) means that their motions are undetectable by the unaided eye in a lifetime.
   - **B.** Their brightnesses are determined by the energy they emit and their distances from Earth; both change by less than 0.1% in a human lifetime for stars in equilibrium.
   - **C.** Their colors are determined by their surface temperature; this is also constant as long as a star is in balance.
   - **D.** Occasionally, certain stars are observed to change.
     1. Some stars have violent magnetic storms on their surfaces, leading to giant flares that temporarily brighten the star.
     2. Some stars, obviously not in equilibrium, pulsate regularly, with periods from days to years, changing rhythmically both their color and brightness.

3. Rarely, a new star will suddenly appear, brightening by a factor of 10 billion and going from undetectability to being the brightest object in the sky—which certainly suggests instability.

   E. Both the overwhelming constancy and the rare exceptions suggest that stars are both long-lived and evolving.

II. Stellar lifetimes are so long that their evolutionary pattern must be inferred from studying many stars in different stages rather than watching a single star evolve.

   A. An alien visiting Earth for only 1 week might be able to infer changes taking place throughout a human lifetime if she collected enough of the right kind of data.
   1. In making physical measurements on a large number of subjects, patterns would emerge relating quantities such as height and weight.
   2. The number of subjects in each stage is proportional to the time spent in that stage.
   3. Collecting data in a playground at a school in which successive ages came out for recess would provide additional hints about growth patterns.

   B. For stars, we act like the alien.
   1. We collect temperatures, brightnesses, radii, masses, distances, and chemical compositions for millions of stars.
   2. Our analogy to the schoolyard is to study star clusters, collections of stars born out of a single cloud of material at the same time.

   C. Our most useful diagram plots surface temperature versus stellar luminosity, the amount of energy emitted each second.
   1. This Hertzsprung-Russell (H-R) diagram is the key to unlocking the story of stellar evolution.
   2. For temperatures of 3000 K to 50,000 K, more than 90% of stars have luminosities that increase systematically with temperature, ranging from 0.1% of the Sun's luminosity at the low-temperature end to nearly a million times the luminosity of the Sun for the hottest stars.
   3. This "main sequence" of stars also has solar mass (M_o) increasing systematically from coolest to hottest: 0.08 M_o at 3000 K to 80 M_o at 50,000 K.

4. From cluster studies and stellar models, we infer the main sequence is also a sequence of lifetimes: from 3 million years at the high-mass, high-temperature end to more than 100 billion years at the low-mass, low-temperature end.
5. A small fraction of all stars do not lie on the main sequence; these provide a key to the various stages of stellar evolution.

III. The story of a star's life can be interpreted from the sequence of locations it occupies on the H-R diagram.
   A. Stars begin life as large, cool, collapsing clouds off to the upper right side of the H-R diagram.
   B. As they collapse, making conditions in their cores hot and dense enough to turn on nuclear reactions, they settle onto the main sequence.
   C. After hydrogen in the core is exhausted, the reactions cease.
      1. The outflow of energy stops and the core begins to collapse, getting hotter and denser.
      2. In compensation, the outer parts of the star expand and cool, sending the star up and to the right in the H-R diagram.
      3. When the core is hot and dense enough, helium can be transformed to carbon, energy flows outward again, and the star reaches a new equilibrium.
      4. This red giant phase continues until the core becomes pure carbon; further readjustment is then required.

IV. Mass is the key parameter of a star, determining its location on the main sequence, specifying the new elements it can produce, and predicting its fate.
   A. The Sun will first burn the hydrogen in its core, then the hydrogen in a shell around the core, then the helium in the core, then the helium in a shell.
      1. Its nuclear-reacting core will shrink to the size of the Earth (1% of the Sun's current radius).
      2. The atmosphere will expand by a factor of 200, until the Earth is orbiting through its outer fringes.
      3. In the final, unstable phases of helium shell burning, the Sun will start to oscillate until the outer layers are puffed off and roughly 40% of the mass is lost.

4. Some of this material will be the helium and carbon cooked up in the core; thus the ejected material will enrich the clouds out of which new generations of stars will form.
5. The Earth-sized core will be revealed as a hot (>100,000 K), tiny (6000 km radius), dead star with no means of generating energy; it is called a white dwarf.

B. All stars with initial masses less than 8 M_o end up as white dwarfs.
1. However, those with greater masses than the Sun can do multiple core-burning cycles and make more kinds of elements.
2. The most massive ones end up as white dwarfs made of oxygen, neon, and magnesium, although some of these, and all the lighter elements, also enrich the star's ejecta.
3. The maximum possible mass for a white dwarf is 1.4 M_o, so the most massive of these stars lose more than 80% of their mass before dying.

C. Stars greater than 8 M_o lead much more dramatic lives.
1. Despite their much larger fuel supplies, they use their supplies up much faster, living much shorter lives.
2. Their much larger masses produce higher pressures and temperatures in their cores, leading to the fusion of elements all the way up to number 26, iron.
3. Recall from Lecture Eighteen that iron is the most stable configuration of nuclear particles.
4. When the star tries to make cobalt and nickel (atomic numbers 27 and 28), energy is absorbed, rather than emitted, and the core suddenly collapses.
5. The energy released blows the star apart, spewing elements 1–26 into interstellar space and synthesizing elements 27–94 (at least) with the energy of the explosion.
6. The collapsed core crushes electrons into protons, making neutrons—thus, a "neutron star" is born, such as the one that may have irradiated the young solar system.

D. The relative abundances of all the elements in the galaxy today are in accord with this model of the stellar synthesis of all atoms heavier than helium.

V. We have completed a large fraction of our journey by understanding the process of stellar nucleosynthesis and thereby identifying the origin of the atomic constituents of our galaxy, the Milky Way.

A. The oldest stars in the galaxy have only $\frac{1}{10,000}$ of the heavy elements that are present in our Sun, and stars being born today are richer in such material by a factor of 2–3—all as a result of the steady creation of new atoms in stars.
B. This pushes the question of origins back a step. Where did the original hydrogen and helium atoms come from that constituted the primordial material of the Milky Way?
   1. To answer this question, we must move far beyond the confines of our galaxy.
   2. And to understand how we read this truly ancient history requires forging a link between space and time.

Suggested Reading:

Bennett, Donahue, Schneider, and Voit, *The Essential Cosmic Perspective.*

Questions to Consider:

1. The maximum mass of stars is shrinking as the universe ages. How will this affect the chemical evolution of the cosmos?
2. Do you really believe that the atoms in that potato chip you are about to eat used to be inside a massive star?

Lecture Nineteen—Transcript
The Lives of Big Stars—Cooking Up Big Atoms

Nearly all the stars you can see on a clear, moonless night are in hydrostatic equilibrium, that perfect balance between gravity and thermal pressure, the latter maintained against the loss of energy from the surface by the nuclear reactions in the core of the star. In the case of our Sun, 600 million tons of hydrogen is transformed to 596 million tons of helium each second, with the remaining 4 million tons of mass turning into pure energy. Four million tons is enough to fill a coal train from New York to Philadelphia, and each second, that much mass disappears into energy—1 coal train gone, 2 coal trains gone, 3 coal trains gone, every second throughout the Sun's history.

This process is irreversible, and eventually, it must end, when the core of the Sun becomes pure helium. Then, change is inevitable. This is the story of stellar evolution and the rich world of atoms it creates. It carries our story back yet further in time, before the formation of the Earth, even before the formation of the matter that made the Earth and all of those isotopes—the beryllium, the carbon, the nitrogen, oxygen, aluminum, and argon, iodine, iridium, rubidium, and strontium, plutonium, thorium, and uranium—that has made our reconstruction of Earth's history possible.

When you go outside and look up at the night sky in a clear, dark place far from city lights, you see a pattern of stars. It may look like there are millions, but there are really only about 1000 or maybe 2000 that you can see with your naked eye. And if you come back the next night, you'll find that all of those stars are in precisely the same place, shining with precisely the same relative brightness, and, if you manage to notice the subtle differences, having all the same colors. And, indeed, if you come back 50 years later and look at that same patch of sky, you'll again see the identical stars in their identical locations with their identical brightnesses and identical colors. This leaves you with a misimpression of the universe, that it's a static and unchanging place.

Nothing could be further from the truth. Stars are born, live out their lives, encounter a rather boring middle age—which some of us understand—and then die in a process, in fact, that's considerably better understood than the human lifetime. I can predict to a fraction of a percent when any given star will die. I can't make that prediction about humans, fortunately. Stars are, in fact, moving through the sky, and they're changing their brightness and their colors. They just do so on a long, long timescale.

The stars are moving, but their enormous distances—the closest one being 26 trillion miles away—means that their motions are completely undetectable with the unaided eye in a whole human lifetime. Their brightnesses are determined by the energy they emit and their distances from Earth. Farther away means fainter for any given amount of energy emitted. But both of these, the distance and the brightness, change less than a fraction of a percent in a human lifetime, for stars that are in equilibrium, at least. Their colors are determined by their surface temperature; this is also constant as long as the star is in balance.

Occasionally, certain stars are observed to change. Some stars have violent magnetic storms on their surfaces—like our Sun but a thousand times more intense—leading to giant flares that temporarily brighten the entire star. Some other stars, clearly not in equilibrium, pulsate regularly with periods ranging from days to years, changing rhythmically both their color and their brightness as they get smaller and bigger, hotter and cooler. Rarely, a new star will suddenly appear, brightening by a factor of 10 billion, going from complete undetectability even with a large telescope, to being the brightest object in the sky. That certainly suggests instability. It's the combination of the overwhelming constancy of the vast majority of stars and the rare exceptions which suggests stars are both long-lived and evolving.

Stellar lifetimes are so long that their evolutionary pattern must be inferred from studying many stars in different stages of life rather than watching a single star move through its life. Imagine an alien visiting the Earth. She only has a week to gather all the data she needs to understand the changes that take place over a human lifetime. How would you even approach such a problem? Clearly, from one week to the next, you or I don't change in any noticeable way. And yet I certainly looked different 50 years ago than I do today, and in another 50 years, I won't look like this at all.

What the alien would do is what any good scientist would do, and that's just gather as much data as possible in the limited time that she has available. It's not even obvious what data to gather, and she probably would be too shy to go up and make measurement of your waist size or hat size, but she can easily, perhaps, determine height, weight, and let's say, hair color. Are those relevant to the evolution of humans? Well, a priori she wouldn't know. But in making a large number of measurements on a large number of subjects, patterns might emerge.

For example, once she got back in her spacecraft and was off to her next destination, she might start plotting the data, like scientists do, one quantity versus another, to see if a pattern emerges. Perhaps on the x-axis she'd plot weight and on the y-axis she'd plot hair color ranging from platinum blonde to black and, maybe, a few places in the East Village, blue. What does she see? The points scatter everywhere. There are very low-mass people that have black hair and very high-mass people that have black hair. There are very low-mass people with blonde hair and high-mass people with blonde hair. The points scatter, randomly, all over the diagram. The blue ones, well, they tend to be isolated in some skinny youth and some aging small old people, so they might be clustered a little bit. But, in general, hair color versus weight doesn't reveal much of a pattern.

On the other hand, replacing height on the y-axis from hair color, one would see a very interesting pattern of points emerging. There are 2 things to notice. First of all, there's a general correlation between weight and height. You might say, well, that's not surprising, but it's not obviously the case. The correlation does not imply causation, so the alien cannot infer that one starts in the lower left of the diagram and evolves towards the upper right. Indeed, my mother-in-law is actually shrinking, so she's evolving from the upper right to the lower left.

The other thing to notice about the diagram, however, is the density of points. There are very few points at low heights and low weights and there are very few points at very high heights and high weights. Why is that? Two different reasons. There are very few points at low weights and low heights because we don't spend very long in that state. We start growing as soon as we're born, and we quickly leave that state and move towards the middle of the diagram. There are very few points at high [heights] and high weights because very few of those kinds of individuals are formed. Nonetheless, the density of points throughout the diagram can tell us something about where humans, in this case, spend most of their time, because the number of subjects in each stage is proportional to the time spent in that stage. This harkens back to the museum analogy I used last time; the number of people in any room in a museum is proportional to the time that they spend in that museum.

Collecting data in a playground at a school, in which successive ages of students come out for recess, would provide additional information about these growth patterns. You'd find a scattering of points when the first graders came out that was low and to the left on the diagram. And as successive grades came out, the points would gradually move up the diagram. This might tell you something about how humans evolve.

For stars, we act just like the alien. We start by collecting lots of data: temperatures, brightnesses, radii, masses, distances, and chemical composition for literally millions of stars. Our analogy to the schoolyard is to study star clusters, collections of stars that are born out of a single cloud of material (so they're all born of the same stuff), at roughly the same time. Because as these clouds collapse, they fragment and tens to hundreds to even thousands of stars can be born in an astronomically short period of time.

For our most useful diagram, we plot points of one quantity versus another. And the one that emerges as we're viewing the most information is when we plot temperature versus luminosity, or the amount of energy a star emits, which we can infer from how bright it appears and how far away it is. This diagram is shown here and is called the Hertzsprung-Russell diagram. Russell is the same fellow who first dated the Earth, using uranium, at 4 billion years, and Hertzsprung is a Danish astronomer who independently discovered these same relationships. This diagram is the key to unlocking the story of stellar evolution. Again, it plots on the y-axis the luminosity or energy emitted per second in the star and on the x-axis plots the temperature. But for historical reasons, plots it backwards, with low temperatures on the right-hand side and high temperatures on the left-hand side. When this diagram was first constructed, people had no idea of its physical meaning; they were just plotting colors of stars, and plotting blue to red was just as sensible as plotting red to blue.

The temperature scale ranges from 3000° for the coolest stars to over 50,000° for the hottest, bluest stars—again, on the left of the diagram. Over 90% of the stars with this range of temperatures have luminosities which systematically increase with temperature, ranging from less than $\frac{1}{10}$% for the coolest stars, in the lower right of the diagram, to nearly a million times the luminosity of our Sun for the very hottest, bluest stars on the upper left of the diagram.

We call this string of stars, from lower right to upper left, the main sequence, because it's where most stars, more than 90%, are found. It turns out when we look at the mass along this sequence of stars, we find that it perfectly correlates with the temperature. That is, the lowest-mass stars are found at the lower right—3000° stars have masses of 0.08 times the mass of the Sun, 8% of the stellar mass being the minimum amount that can ignite nuclear reactions at the center, as I mentioned last time. And going up the diagram, we come first to $\frac{1}{2}$ solar-mass stars, and then the Sun, and then stars twice as massive, 10 times as massive, and ultimately, at 50,000°, in the upper left, stars that are 80 times the mass of the Sun, the maximum mass that a star can have and live in equilibrium.

From cluster studies and from stellar models, we infer that the main sequence is also a sequence of lifetimes. Now, you might think that the most massive stars have the most fuel and would live the longest. That would be true if you failed to notice that the most massive stars also shine the brightest. They produce vastly more energy each second than the lower-mass stars, and that takes more fuel. So, in fact, the opposite is the case. The stars in the upper left, the 80 solar-mass, 50,000° stars, only live for 3 million years—million with an M, which in astronomy is a very short time. Whereas the low-mass stars, with a modest amount of fuel, use it even more modestly—down at 3000°, less than $\frac{1}{10}$% of the luminosity of the Sun—and they live for hundreds of billions of years. Our Sun, roughly in the middle of the diagram with its 1 solar mass of material, uses it at a rate that will last for 10 billion years. A small fraction of stars do not lie on the main sequence, and these provide a key to the various stages of evolution. The stories of a star's life can be interpreted from the sequence of locations it occupies on the H-R diagram. Stars begin, as I have said, as large, cool clouds of gas; "cool" means off to the right-hand side of the diagram, even below 3000°. And if they're large, they can radiate a fair amount of energy, albeit in the infrared part of the spectrum, and so they lie over here. As they begin to collapse, they shrink down, the core gets hotter, the outside gets hotter, so they move to the left in the diagram. As they shrink, they get smaller and, as a consequence, have less emitting area and emit less light, and so they fall down in the diagram. They descend, as we say, to the main sequence. And they adopt a location that's appropriate for their particular mass.

As they collapse, they eventually make conditions in their core hot and dense enough to turn on the nuclear reactions. And then equilibrium is set, gravity is balanced by thermal pressure, and the star lands on the main sequence, a location it will hold for most of the rest of its lifetime, the boring middle age I referred to before.

This goes on until no hydrogen is left in the core. The outflow of energy that has been produced by this continuous furnace of nuclear reactions—protons fusing with protons, deuterium fusing with protons, helium fusing with helium to make helium-4—this energy flow will stop when hydrogens cease to exist. And the core, having lost its source of support, the energy flowing out from these reactions, begins to collapse. In collapsing, 2 things happen: The same amount of material compressed into a smaller space means it gets denser, and as matter falls down, it gets hotter, because gravitational energy is turned into the thermal energy of the particles running around inside the star. In compensation, it turns out, the outer parts of the star—which are not participating in these nuclear reactions; only the inner 10% have conditions where fusion can occur—the outer part, sensing this extra energy from the hotter core of the star, begins to expand and cool, sending the star in the H-R diagram up, because it's getting bigger and radiating more energy, and to the right, because it's getting cooler. This is the realm of the so-called red giants.

When the core is hot enough and dense enough, helium can be transformed into carbon, as I noted last time, by having 3 nuclei of helium simultaneously collide with each other and snap together to form the very, very stable carbon nucleus. Energy, then, is flowing outward again, and the star adjusts to reach a new equilibrium. The outer part is still much bigger than it was before, and the star on the outside is cooler than it was before. The helium reactions go even faster than the hydrogen reactions and, therefore, generate even more energy as they burn helium to carbon, but that energy is spread out over a larger surface, and therefore, the temperature is lower, although the luminosity, the energy emitted each second, is higher. This red giant phase continues until the core becomes pure carbon, and then further readjustment is required.

Mass, it turns out, is the key parameter of a star. It determines its location, where it lives on the main sequence for most of its life; it specifies which new elements the star can make; and it predicts the star's ultimate fate. The Sun will first burn hydrogen in its core; then in a shell around its core, when the core hydrogen is exhausted, but a thin layer around it might be hot enough for fusion to occur; then, when all the hydrogen is exhausted and the outside is expanded and the core has contracted, the temperature and density get high enough for helium to bind into carbon in the core; and then ultimately, when that's all exhausted, helium in a shell.

The nuclear-reacting core will shrink at that point because there's nothing to support it against gravitational collapse. And it will shrink down to the size of the Earth, 1% of the Sun's current radius. The atmosphere will expand by a factor of 200 until the Earth is orbiting through its outer fringes, not a very pleasant prospect for life or even the oceans on Earth, which will be instantly evaporated. In the final unstable phases of helium burning, where the nuclear reactions are very, very sensitive to the density and temperature available, helium will ignite in a shell around the core of the star. And that'll be a big pulse of energy, and so the star will start to expand because it has been kicked from the inside. As it expands, the density goes down and the temperature goes down, and that's enough to shut the reactions off.

Then, there's no support coming out of the center of the star and the star starts to collapse again. But as it collapses, the temperature goes up and the density goes up, so the nuclear reactions turn on again. It reaches this unstable phase, where it's turning on and turning off, and turning on and turning off, and turning on and turning off, until the outer layers of the star get too big a kick and are puffed off and go drifting off into space. Over the course of many of these cycles, up to 40% of the Sun's mass will be lost. Some of this material will be the helium and the carbon cooked up in the core of the star before its demise. Thus, this ejected material will enrich the clouds out of which new generations of stars will form, in particular, seeding them with the crucial element carbon—crucial, that is, for life.

The Earth-sized core will, when the final puff occurs, be revealed, as the outer layers of gas drift off into space. It is a hot (more than 100,000°), tiny (only 6000 kilometers in radius, the size of the Earth), dead star, where by "dead," I mean a star which has no means of generating energy. The only energy it radiates, the only reason we can see it at all, is because it's hot. It's in Kelvin's model; the star formed hot, radiating away its energy. The star is called white dwarf. It's very dense, 1 teaspoonful of this material weighs a ton—so imagine your SUV compressed so it fits in a teaspoon—but it's dead because it has no way to generate energy other than to cool off.

There are billions of these stellar corpses scattered around our galaxy. Their age is easily determined by their temperature, because all they do is cool. In the lifetime of our galaxy, nearly 10 billion years, the coolest ones have gotten down to about 5000° or so, so they're still visible, albeit faint because they're so tiny. And as time goes forward, they'll just cool further, and further, and further, until they fade away as dead lumps of material, never to participate again in the cycle of life and death in stars.

All the stars with initial masses that are less than about 8 times the mass of our Sun end up as white dwarfs. However, those considerably more massive than the Sun—4, 5, 6 times the mass of the Sun—can do multiple core-burning cycles before they die and, thus, make more elements. They begin, as all stars do, by turning hydrogen into helium in the core. Then, when the core is pure helium, the equilibrium readjusts so the temperature and density are higher and helium can fuse to make carbon. That's the end of the process for our Sun, but for these more massive stars, with lots more material overlying, squeezing on the center of the star, the next readjustment can make the density and temperature high enough that the carbon can start burning and make oxygen, and neon, and other elements. And then, when the whole core is turned to oxygen and neon, then it'll readjust again, get hotter and denser still, and oxygen and neon can turn into magnesium, and silicon, and sulfur, and other elements, as well.

For stars less than 8 times the mass of the Sun, these cores, then, are made up of oxygen, magnesium, and neon, in particular, 3 of the more common of the heavy elements, as well as a lot of the other lighter elements. And when their outer envelopes get puffed away, all of those new elements get distributed to interstellar space.

It turns out the maximum possible mass for a white dwarf is 1.4 times the mass of the Sun, 40% more massive than the Sun. So these stars that start out as nearly 8 times the mass of the Sun need to lose over 80% of their mass in these unstable, dying phases before their cores can shrink to become white dwarfs.

Stars that start out life with more than 8 times the mass of the Sun lead much more dramatic lives. Despite their much larger fuel supply, they use it up much faster. They're profligate, living much shorter lives, as I've said before. Their much larger masses produce much higher pressures and temperatures in their cores, leading to the fusion of elements all the way from hydrogen to helium and helium to carbon, to oxygen, neon, magnesium, silicon, sulfur, and all the way to element number 26, iron.

You recall from Lecture Eighteen that iron is the most stable configuration of nuclear particles, 26 protons and 30 neutrons nestled together, attracted by the strong nuclear force in the most stable configuration that nature has devised. Thus, in the process of building up these elements, from hydrogen to helium to carbon, etc., we're moving down the binding energy curve. And each time we take a step down, just like when you're walking down the stairs, you give off energy. And so all of these processes up to iron give off energy, help hold the star up, and allow the star to shine.

However, when the core turns to iron from silicon—a step that only takes a week, one of our weeks; the entire solar mass and a half of material at the center of a massive star turns from silicon to iron—it then tries to do the next logical thing. Once the core is all iron, well, the core has to collapse again, raise the temperature a little bit, raise the density a little bit, and start squeezing some of those particles together to make elements 27 and 28, cobalt and nickel. Why not? It's what the star has done successfully 26 times before. But since iron exists at the bottom of the nuclear binding energy curve, moving away from there requires energy.

When you're at the bottom of a valley, to get out, you have to climb, and climbing absorbs energy rather than giving it off. And so when the nuclear particles in the core try to add a few protons or neutrons to iron nuclei, rather than giving off energy, as this process of fusion has done in the past, all of a sudden, energy is absorbed, because the iron has to move up the binding energy curve. The nuclear reaction, which throughout the whole life of the star has acted as a counterbalance to gravity, which was always trying to bring the star in, suddenly sucks energy out of the center of the star, aiding gravity and, gravity being very patient, crushes the star.

The star goes from a size roughly the size of the Earth to something roughly the size of Manhattan. It shrinks from 6000 kilometers in radius to 10 kilometers in radius, and it does so in half a second. That's an enormous amount of gravitational energy released in an incredibly short period of time. And a lot of energy released at one moment in time in a particular place is usually called an explosion. And that's exactly what occurs.

energy released in that final step of gravitational collapse blows the star to smithereens, spewing out all the elements, numbers 1 to 26, that have been cooked up in the star over its lifetime into interstellar space. And, because there's so much extra energy available in this enormous explosion, synthesizing all of the other elements, from number 27 to number 92 and beyond. The energy of the explosion slams nuclear particles together willy-nilly, creating all the different isotopes of all the elements in the periodic table in that brief moment of stellar explosion. The synthesis of the elements is a long, slow process in the core of a star for numbers 1 through 26 but an instantaneous one for all the other elements. It's not surprising, then, that those other elements are rare, as they only occur in these relatively rare exploding stars we call supernovae.

The star blows up, except for this inner 1.4 solar masses of iron or so, which collapses in on itself with such a force that the electrons are driven into the protons of the nucleus. Electron plus proton, negative plus positive, turns into neutral or a neutron. The neutrons that are there stay behind, and an entire star, a ball of neutrons, is born. This is a star such as the one that may have irradiated our young solar system and given us our left-handed molecules for life.

Neutron stars are the most extreme state of matter known. It is an entire star with the density of an atomic nucleus. A single teaspoonful here weighs a billion tons. Imagine all the cars, and trucks, and SUVs on the planet compressed into the size of a sugar cube. That's what a neutron star is.

The relative abundance of the 92 elements in the galaxy today is actually in accord with this model, in which stars synthesize all of the atoms heavier than helium. In fact, it's even better than that; the abundance ratios of uranium to carbon or oxygen to neon are predicted by this model of atoms being synthesized in the cores of stars up to iron and in the explosions which end the lives of massive stars for all the rest.

As the late astronomer Carl Sagan was fond of saying, "We are star stuff." It is literally true. I have said more than once that you are what you eat, but all that you eat, excepting only the hydrogen atoms, was cooked for you inside a star.

Understanding the lives of stars takes us another step back along our journey, identifying the origin of the atomic constituents of our galaxy, the Milky Way. We do this by understanding the process of stellar nucleosynthesis, the creation of nuclei in the cores of stars. The oldest stars in our galaxy have only $\frac{1}{10,000}$ of the heavy elements that are present in our Sun. Stars being born today are richer in such material by a factor of 2 or 3, all as a result of the steady creation of new atoms as generations of stars go by.

Understanding this, however, just pushes the question of origins back 1 step further. Where did the original hydrogen and helium atoms come from that constituted the primordial material out of which the Milky Way was formed? To answer this question, we must move far beyond the confines of our own galaxy. And to understand how we read this truly ancient history requires forging a link between space and time. That will be our task next time.

Lecture Twenty
Relativity—Space and Time Become Spacetime

Scope:

To complete our quest—to understand the origins of the very constituents of our atomic historians—we must explore the first moments of the universe. How is this possible? By just looking. As a consequence of the finite speed of light, looking out into space is precisely equivalent to looking back in time. It takes light 8.3 minutes to get from the Sun to the Earth. Thus when we "see" the Sun, we are not seeing it as it is "now" but as it was 8.3 minutes ago. The consequences of this small delay are insignificant. However, if we wish to see what the universe was like 6 billion years ago, all we need to do is look out 6 billion light-years into space and the picture is laid out before us—again, we see "then," not "now." This unavoidable mixing of space and time has profound implications, as Einstein was the first to recognize in his theory of relativity.

Outline

I. Light travels at a finite speed, which means it takes a finite and calculable amount of time to cover a given distance.
   A. Since the speed of light is 300,000 kilometers (186,000 miles) per second, in our everyday lives it appears instantaneous.
      1. One gauges the distance of a thunderstorm by counting the seconds between the lightning strike and the roll of thunder, since sounds travels so much more slowly.
      2. Old-time intercontinental satellite phone links gave a hint of the problem—the signal traveling 23,000 miles up to the satellite and then 23,000 miles back down to Earth led to an echo with a few tenths of a second delay.
   B. Once off the Earth, however, this delay starts to get significant.
      1. Astronauts on the Moon were 2.6 seconds away from instructions.
      2. The rovers on Mars take 10–40 minutes for a round-trip "conversation."
      3. The nearest star is 4.4 light-years away—which means we see it not as it is today but as it was when the light left it on the journey to Earth 4.4 years ago.

4. Symmetrically, looking back on Earth from a distant star, one sees the past, not the present.
5. The center of the Milky Way is about 25,000 light-years away; a big telescope there would see a very different Earth.
6. However, 25,000 years is a short time in terms of stellar lifetimes, so no big changes are expected in how stars look there.

C. Outside of the galaxy, the time scales get longer quickly.
1. The nearest large galaxy, Andromeda, is 2 million light-years from Earth; a civilization studying Earth from there now would see no humans.
2. The nearest rich cluster of galaxies is 100 million light-years away; astronomers there would today see dinosaurs roaming the Earth.
3. From 2 or 3 billion light-years away, the universe looks noticeably different.
4. The most distant objects detected at 12.5 billion light-years are in a radically different-looking universe that is, as we see it today, only 1 billion years old.
5. But of course out there it is "today" too—we have to think carefully about what we mean by "now," and that leads us to thinking about the connectedness of space and time.

II. "Nothing puzzles me more than space and time, yet nothing troubles me less," said Charles Lamb.

A. The difference between an essayist and a scientist is that if something puzzles one, it also troubles one.
1. From Maxwell's understanding of the link between light and electromagnetism in the 1860s onward, troubling thoughts about space and time were brewing.
2. Fitzgerald, Lorentz, and Hertz recognized inconsistencies in Maxwell's model.
3. Michelson and Morley's experimental failure to find the "aether"—the substance through which Maxwell's light waves supposedly traveled—further clouded the issue of what electromagnetic waves really were.
4. In 1905, Einstein crystallized the problem in a paper entitled "On the Electrodynamics of Moving Bodies."

B. In Einstein's relativity, the seemingly separate concepts of space and time become linked—through the special status of light—into spacetime.

C. The first postulate of relativity is that absolute uniform (constant velocity) motion cannot be detected.
   1. You may think it is obvious that if you are standing watching a train go by, the train is moving and you are standing still.
   2. However, you are, in fact, moving—at 30 km/s around the Sun, which is moving at 230 km/s around the galaxy, which is moving at nearly 600 km/s through spacetime—so who is standing still?
   3. Billiard players don't change their shots from spring to fall, but the difference in their motion through space is 216,000 km/h.
   4. You judge motion "with respect to" something else; i.e., motion is relative, and there is no experiment you can perform to detect constant-velocity motion.

D. The second postulate of relativity is that the velocity of light in a vacuum is constant, independent of the motion of the source or the observer.
   1. Rocks do not obey this postulate; a rock thrown at you from a car driving toward you has a higher velocity than a rock thrown by a stationary adversary.
   2. Sound, a wave like light, obeys this postulate, but only in a single special reference frame—the air through which it propagates.
   3. Since the Michelson-Morley experiment established that light has no medium through which it propagates—it travels through the pure vacuum of space—Einstein's postulate is reasonable: Its speed is the same in all reference frames.

E. Following the logical consequences of these 2 postulates leads to a picture of the world in which space and time are mixed together (as spacetime), to some bizarre, but experimentally verified, consequences.

Suggested Reading:

Bias, *Very Special Relativity*.

Einstein, *Relativity*.

Taylor and Wheeler, *Spacetime Physics*.

Questions to Consider:

1. What are some examples other than a thunderstorm in which you can notice the difference between the speeds of light and sound?
2. How can you tell you are moving at night in a plane?

Lecture Twenty—Transcript
Relativity—Space and Time Become Spacetime

To search for the origin of the primordial elements and, even more fundamentally, their constituent subatomic particles, we need to travel back to near the beginning of time. This may seem like an impossible task, but the finite speed of light comes to our aid, because looking out into space is looking back in time.

Light travels at a finite speed, which means it takes a finite and calculable amount of time to cover a given distance. The speed of light is 300,000 kilometers per second, or for those of you still stuck with English units, 186,000 miles per second. That's enormously fast, and so in our everyday lives, it appears instantaneous. As you probably know, you can gauge the distance of a thunderstorm by watching the lightning strike and then counting 1-one thousand, 2-one thousand, 3-one thousand, 4-one thousand, 5-one thousand, and hearing the clap of thunder. Sound travels through air at about 300 meters per second; the light gets to you at 300,000 kilometers per second, so it, of course, gets to you effectively instantaneously. And if the thunderstorm is 1500 meters or 1 mile away, it takes 5 seconds for the sound to catch up.

Old-time intercontinental satellite phones, some of you may remember, give a hint of this problem. They sent a signal from where you were calling to a satellite 23,000 miles above the Earth, because that distance, 23,000 miles above the surface, is the point a satellite has to be at if it is to orbit around the Earth at the same rate that the Earth rotates; thus, it stays above the same point on Earth all the time, ready to receive your signal. So the signal travels up 23,000 miles and then down 23,000 miles. And when you talked on such phones you could hear a very annoying echo a couple of tenths of a second after you made a sound, because you were hearing the reflection off the satellite where the total signal had traveled over 46,000 miles.

Once one gets off the Earth, however, this delay starts to get significant. As some of you probably remember and others may have been told, astronauts did actually walk on the Moon in the late 1960s. Probably some of you remember that 2:00 a.m. event in July of 1969 when Neil Armstrong made that small step for man and that great leap for mankind. That event was seen around the world by nearly a billion people, and that's when there were only 3 billion people on the Earth all together. It was a momentous event. But

NASA kept at it. They kept sending astronauts back to the Moon. They went to different parts of the Moon. They picked up different Moon rocks, which were useful in the story I told you earlier about the origin of the Moon. But the public started to lose interest.

So they thought: What can we do to make the public really want to watch these Moon shots again? What do people really like to watch on TV? And they came up with the brilliant idea of golf. Now, I must confess, why people like to watch golf on TV has never been clear to me, but clearly, they do. And they thought: Ah, what we can do is have one of the astronauts hit a golf ball on the Moon. In the weakened gravity of the Moon, it would be easy to drive a ball say a mile or so and people would be really impressed. And so they set up to go to the Moon again, carrying a golf club, a tee, and a golf ball.

The first astronaut came down the ladder and set up the camera, connected the cables, and got ready for the shot. The second astronaut came down with his golf club and in the process hooked his boot under the cable between the camera and the satellite. The controller in Houston, seeing this on his screen, of course, immediately shouted into his communications device, "Look out! You're going to trip on the cable." And then, of course, the screen went blank. Because what the controller saw in Houston, 380,000 kilometers away, was the picture of the boot hooked under the cable 1.3 seconds ago. And his command to the astronaut to avoid tripping took another 1.3 seconds to get there; 2.6 seconds is plenty of time for the cable to have been ripped out and the golf game to have been delayed.

If we go farther out into the solar system, this problem gets even more acute. The little rovers that have traveled around on the surface of Mars for some years are between 10 and 40 minutes—depending on the relative position of Earth and Mars in their orbits around the Sun—away. It takes a roundtrip conversation between 10 and 40 minutes. And so you can't have a controller sitting at JPL in California, watching the little camera on the head of the rover as it moves across the surface, and say, "Oh look, the rover is nearing the edge of a crater." Because by that time, 10 minutes later, you have a billion dollars of space junk at the foot of the crater. You have to move very carefully when conversations take half an hour.

The nearest star is 4.4 light-years away, which means we see it not as it is today, but as it was when light left it on its journey to Earth 4.4 years ago. There's nothing we can do to see what that star looks like today, in our today, because the light takes 4.4 years to get to us. Symmetrically, looking back at Earth from that star, if you lived on a planet that orbited that star, one sees the past, not the present. Let me illustrate.

Take the star Altair, one of the bright stars you can see with your naked eye. It's about 17 light-years away. If you were an astronomer on a planet orbiting Altair and you looked back at Earth, what would you see? No gray hair, lots more hair; not bad. You'd see Earth as it was 17 years ago. If you were on a planet orbiting Arcturus, another bright naked-eye star that's 39 light-years away, and looked back at Earth, you'd see this. Not even a beard, and note the little National Honor Society pin there on the lapel, very sexy. If you were on Capella, 53 years light-years away, and looked back at Earth, you'd see this absolutely adorable platinum-haired young man. And if you were on Aldebaran, which is 56 light-years away, looking back at Earth, you'd see the Michelin tire baby.

Going even father out into space, this starts to have serious consequences. This is an image of the constellation Orion, one of the few constellations that I can actually recognize. It's got those 3 bright stars in a row that form the belt of the constellation which the Greeks thought of as the hunter. Just below the belt, the little smudge, represents that region of star formation I talked about, the nearest big stellar factory to Earth. But the bright red star at the head of Orion is about 400 light-years from Earth and is called Betelgeuse.

This is a very interesting star. It's a red super giant and it's a star very near the end of its life. It has exhausted most of its nuclear fuel and is about to explode. When I say "about to explode," we can't predict that to better than maybe 10,000 or 20,000 years. It could, in fact, have already exploded. Indeed, it could have exploded about 399 years and 364 days ago. There's no way I could know that. But when I go outside tonight and look up at the sky, I could see a star that's a billion times brighter completely dominating the night sky, as the evidence of that explosion finally, 400 years later, arrives at Earth. Incidentally, this is what I looked like 400 years ago when I posed for Raphael.

The center of the Milky Way is about 35,000 light-years away. A big telescope there would see a very different Earth. Indeed, one would see the Neanderthals wandering around Europe and making pretty cave paintings of bison. However, 35,000 years is a short time in terms of stellar lifetimes, as we've learned, so there would be no big changes expected in how the stars look there, and indeed, we don't see stars very differently.

Once we get out of the galaxy, however, the timescales get longer really fast. The nearest large galaxy to us, Andromeda, is 2 million light-years from Earth. A civilization studying Earth from there now would see no humans. The nearest rich cluster of galaxies is 100 million light-years away; astronomers there today would see dinosaurs roaming the Earth. That's their today, of course, but wherever one is, it's your today. Two or 3 billion light-years away, a distance one can see with even a relatively modest-sized telescope, the universe looks noticeably different. Galaxies and stars have evolved on timescales of billions of years. The most distant objects ever detected are about 12.9 billion light-years from Earth, and they exist in a universe radically different from the way the universe looks as we see it today, because their universe is less than 1 billion years old. Very few galaxies have formed. Stars are very young. Heavy elements are almost nonexistent. But, of course, out there it's "today," too. We have to think very carefully about what we mean by "now," and that leads us to thinking about the connectedness of space and time.

The British essayist Charles Lamb wrote at the turn of the 19th century, "Nothing puzzles me more than space and time, yet nothing troubles me less." I guess the difference between an essayist and a scientist: If something puzzles a scientist, it also troubles him deeply. From the middle of the 19th century, when Maxwell wrote down his equations that connect light and electromagnetism, troubling thoughts about space and time were brewing. The Irish physicist George FitzGerald, for example, saw Maxwell's famous 4 equations which unite electricity and magnetism, link the 2 forces together to make light, as an incomplete starting point rather than as a final answer. Hendrik Lorentz and Heinrich Hertz also recognized potential inconsistencies in this model of electromagnetic waves producing light.

On the experimental side, the American physicists Michelson and Morley attempted to measure the medium in which Maxwell's light waves were supposed to propagate. And they repeatedly failed. The point is that a wave, as Maxwell had concluded light was, waves something. Water waves go like this. The molecules of water go up and down as the wave moves forward. Sound waves, as I've described earlier, knock the molecules together like this as the wave propagates by. Something must be waving for a wave to exist. What was it that was waving when a light wave passed through space?

The Michelson-Morley experiment was conceptually very simple, although the technical details were a little daunting. If light really were Maxwell's hypothesized waves, there must be something that was waving. It must occupy all of space, whatever it is, since the light from very distant stars clearly reaches us. In deference to Aristotle's fifth element that was supposed to occupy the celestial sphere, this putative waving substance was dubbed the aether.

Michelson and Morley's idea was as follows: As the Earth completes its annual circuit around the Sun, it must be plowing through this aether. If the aether were stationary—and what "stationary" means will come to be a major part of the problem here, but let's say it's sort of a still ocean and the Earth is moving around through it—then in one season of the year, one should feel the aether "wind" in one's face when one is moving this way, and 6 months later, it should be blowing from one's back as one moves this way. Even if the aether were not stationary with respect to the solar system but were somehow streaming through the solar system, the speed that we detect in it should change as our velocity of 30 kilometers a second, not exactly a slow speed, around the Sun each year continues.

They devised a 2-armed device to test this hypothesis. Light entered from the left of the device, and in moving along, it encountered a half-silvered mirror. That's a mirror that's designed such that roughly half the light that hits it passes right through as though it were glass and the other half of the light that hits it reflects, as from a normal mirror. They set this mirror at a 45° angle, such that the light coming from the left passed through, for some of the waves, all the way over here, and the other of the waves hit at 45°, reflected at the equal angle, and went up.

At each location, the vertically moving light and the horizontally moving light, another mirror was put in place that reflected all the light, and so the light rays started back towards each other again. They then again encountered the half-silvered mirror, and the vertically moving beam passed directly through to a detector down below, whereas the horizontally moving beam reflected off the mirror and also went to the same location where the detector was located. In fact, it was a microscope which they could look in with their eyes.

By carefully adjusting the lengths of the 2 beams to be equal, the light waves should have arrived at the finish line in phase. That is, as the peaks go this way, the same number of peaks should have made it as the peaks that went this way, because the distances were exactly the same. And if the peak lines up with the peak of a wave and the valley lines up with the valley of a wave, we have what we call constructive interference; that is, the total wave gets twice as big, because the peaks add and the valleys add. If, on the other hand, the path lengths are slightly different, say, different by half a wavelength, then the peak of one wave will line up with the valley of another, and the valley of the first wave will line up with the peak of the second, and the wave will actually cancel out. You can see this effect easily in water. So if the waves were traveling through this aether wind that was postulated, they would not arrive simultaneously.

The situation is analogous to 2 swimmers in a fast-running stream. One is asked to swim upstream 100 meters and then back again, and the other is asked to swim across the stream and back again, also 100 meters each way. With a little simple algebra, one can show the transit times for the 2 equally strong swimmers are different. Going up and back against the current takes a different amount of time than going across and back the current. Meaning, for the experiment, that the light would not constructively interfere if there were a medium in which light propagated. Michelson and Morley spent years perfecting this experiment, but however precisely they repeated the test, the evidence for the aether never appeared.

In 1905, Einstein crystallized the problem in a paper he entitled "On the Electrodynamics of Moving Bodies." This is, in fact, the paper that introduced the concept of relativity to the world. In Einstein's relativity, the seemingly separate concepts of space and time become linked together, through the special status of light, into spacetime. Einstein proposed 2 simple postulates. The first postulate of relativity is that absolute uniform motion—constant-velocity motion, that is—cannot be detected. Now, you may think this sounds absurd. It's obvious if you're standing watching a train go by that the train is moving and you are standing still. But, in fact, you're not standing still; you're moving at a pretty high speed. You're moving at 30 kilometers a second, being carried around by the Earth around the Sun. And the Sun and the whole solar system is moving at 230 kilometers a second, around the Milky Way; it takes 250 million years to go around once. And the whole Milky Way, our galaxy, is moving through spacetime at 600 kilometers a second. So who's for you to say that you're standing still and the train is moving?

Nonsense, you say, of course, with respect to the Earth, I'm standing still and the train is moving. Ah, it's "with respect to" something. It's true, of course, that billiard players don't change their shots from spring to fall, even though the difference in their motion through space is about 216,000 kilometers an hour. Nonetheless, the balls bank off the bank and into the pockets in just the same way.

The point is you always judge motion with respect to something else. That is, motion is relative, and what Einstein's postulate says is that there's no experiment you can perform to detect constant velocity, absolute motion. My son discovered this, when he was young, in the New York City subway system. The old subways used to have these long, smooth platforms that ran the whole length of the thing; not individual seats, but the whole length of the car had a long, smooth surface. And when the subway cars were empty, he liked to bring his little toy trucks into the subway and put them on the bench. What he discovered was if he placed the truck in the middle of the bench while the train was rocketing along between stations at 40 miles an hour, the truck sat perfectly still. When the train started to decelerate as it entered the station, the truck would take off in the direction of the train. When the train started to accelerate as it pulled out of the station, the truck would take off in the other direction. Accelerated motion can be easily detected. It's that feeling you get in an elevator when it starts up or the little truck on the subway bench starts to move. But when the train is moving along at a constant speed, there's no experiment you can perform inside that subway car that lets you know that it's moving or indeed how fast it's moving.

The second postulate of relativity is that the velocity of light in a vacuum is constant, independent of the motion of the source or of the observer. Does this make common sense? Rocks don't obey this postulate; for example, a rock thrown at you from a car driving toward you has a higher velocity than a rock thrown by a stationary adversary. Center fielders know this principle well; when they're in center field waiting for a fly ball and a runner's on third base tagging up to run home, they always position themselves many feet behind where the ball is going to land and, as it comes down, start running forward towards the ball, such that when they catch it, their velocity of motion plus their arm's velocity of motion will make the ball reach home plate faster. So physical objects certainly don't obey the rule that their velocities are independent of the motion of the source or the observer. The velocity of the source adds.

Sound is a wave like light, so perhaps we ought to see if this obeys this postulate. It does, but only in a special reference frame, the air through which it propagates. Imagine the following situation: On the Fourth of July every year in New York, the Macy's department store puts on an enormous display of fireworks. They line the East River with barges and thousands of fireworks go up into the sky. A couple of million people usually line the FDR Drive along the East Side of Manhattan to watch this show. Now imagine one year, a dense fog settles over the city; it would certainly put a damper on the fireworks. But you happen to be a friend of the mayor, and he's not going to let the fireworks show go unwatched, and so he orders out his helicopter and takes you up in the helicopter, hovering above the East River maybe 3 kilometers away from the source of the fireworks. The fireworks are launched from the barges, they go up through the layer of fog, they explode in the sky, and you see them and you hear them. You see them essentially instantaneously, because you're only 3 kilometers away and the light travels 300,000 kilometers per second, so it only takes $\frac{1}{100,000}$ of a second for the light to get to you. You can't possibly measure that short a time period; your neural circuits work 100 times more slowly.

You see the light from the fireworks, the beautiful explosion, instantaneously. The sound, however, has to propagate through the air. And you think, in your little physicist way, "Well, I'll measure the speed of sound in the air tonight. Perhaps it's a little different because of the fog." And so you take out your watch, you watch the explosion, and you count and you see how long it takes for the sound to get to you. Since sound travels, as I said, 300 meters per second and you're 3 kilometers away, then it takes 10 seconds. And you go, "Ah, the speed of sound is 300 meters per second; 10 seconds divided into 3000 meters." That's if you're hovering in the helicopter.

Suppose instead that your helicopter pilot decided to fly towards the fireworks to get a better view. In that instance, you would see the flash of light and you would time on your watch how long it took and you would say, "Ah, it only took 7.5 seconds for the sound to get here; sound must travel at 400 meters per second because it has traveled 3000 meters in 7.5 seconds."

Nonsense, you'll say. We were moving in the second case, so of course, we get a different answer, and the answer we get is wrong. But how do you know you were moving? I've been in a helicopter. It's a very noisy and unpleasant experience, but it's very hard to tell when you're standing still or when you're moving if you have your eyes closed and aren't looking out the window—which, I must say, part of the time, I did have my eyes closed. The point is, when you're in an airplane flying from New York to California, you have no sense of motion. When you drop your Coke off the edge of your chair, it doesn't fly to the back of the plane and pin the stewardess against it. It just falls straight down. There's no experiment you can do in the plane or in that helicopter to tell whether or not you're moving at a constant speed. Therefore, there's no way to know whether your measurement of the speed of sound was correct or not.

Ah, you say, yes, I could. I could stick my hand out the window, and I'd feel the air rushing by if we were moving, and then I would know that we were moving. Yes, precisely. The frame of the air is a frame in which sound speed is constant, independent of the motion of the source of the observer, but it's the only thing.

Since the Michelson-Morley experiment established that light has no medium through which it propagates—it travels through the pure, empty vacuum of space—Einstein's postulate is not unreasonable. Its speed is the same in all reference frames.

Following the logical consequences of these 2 postulates leads to a picture of the world in which space and time are mixed together as an entity we call spacetime. It also leads to some bizarre but experimentally verified results. If you find the concept of spacetime confusing, let's take a step back. Just start by trying to define time without any references to time itself: Well, you see, time is, time, time moves through—I mean, the seconds go by—time…. It's not so easy to define time, is it? Time is something you experience. Subjective time may seem to you to run at different rates. In fact, you know, the clock ticks in a regular way, but a clock is just something we've devised to measure time. Is time the same for me as for you? Do my clocks run at the same rate as your clocks run? This is something that we'll have to investigate based on Einstein's 2 relatively simple-sounding postulates.

Science proceeds by constructing logically self-consistent hypotheses and then devising experiments to test their consequences. As I constantly say to my students, although it's not popular with them or necessarily with my colleagues, I don't regard science as a search for Truth with a capital T. I regard science as a search for hypotheses, for models if you will, for the way the world works, which are testable and, most importantly, are falsifiable. If I make a hypothesis, such as Einstein did, I must test it, and there must be a possibility in that test of the hypothesis being proved false. If there's no such possibility, then it's not a scientific hypothesis, because science makes progress not by proving things true but by proving things false. By showing that one's model is incorrect, one makes progress in understanding the universe.

Einstein's relativity was not obviously true. The 2 hypotheses sound simple and innocuous enough: It's not possible to determine if one is moving at a constant speed and that the speed of light, this slightly weird thing, travels at a constant speed no matter who's doing the measurement or how fast the source is moving. The point is that these 2 hypotheses lead to a number of predictions, all of which are testable. And over the last 100 years, Einstein's relativity has passed the tests with flying colors. Indeed, we cannot begin to understand things like the high-energy collisions of cosmic rays with our atmosphere, the source of the carbon-14 atoms we've used so often in this course so far, without applying this radical departure from our commonsense notions of space and time.

Next time, we'll follow the logical consequences of Einstein's 2 postulates through to a new understanding both of how subatomic particles behave and how the universe itself exists in spacetime.

Lecture Twenty-One
(Almost) Everything Is Relative

Scope:

A careful analysis of the consequences of the finite speed of light led Einstein to the realization that the quantities we regard as so different—space and time—are in fact inextricably linked. With respect to a stationary observer, a moving object's length in space actually contracts, while as if in compensation, time is spread out along its length and its clocks run slow. These effects are observed when cosmic nuclei traveling at nearly the speed of light strike the atmosphere and produce new, short-lived particles that should die long before they reach the ground. They are recorded in ground-based detectors nonetheless—time for them has slowed, and their decay is delayed long enough for them to meet Earth's surface. Likewise, the atoms sending light signals from the depths of space, being carried away by the expansion of the universe, signal their distance by the amount by which their oscillations have slowed.

Outline

I. Einstein began with 2 perhaps nonintuitive but straightforward postulates and deduced the logical consequences from them.
   A. This resulted in predictions of bizarre behavior for a body traveling close to the speed of light.
   B. To understand the conflict this engenders with our common-sense notions, we must begin by formalizing the concepts of frames of reference.

II. Imagine me standing by the train tracks watching my wife, Jada, walk through the train with her little dog, Pixel.
   A. The train is moving to the right at 10 m/s from my perspective; from the train it appears that I am moving to the left at 10 m/s.
   B. Pixel is moving in the same direction as the train at 1 m/s; thus it appears to me that Pixel is moving to the right at 11 m/s.

- C. We impose on this picture 2 graphs representing the 2 reference frames corresponding to Jada's view and mine, and we record the coordinates in each.
- D. Common sense leads us to derive the Galilean transformations—a simple set of equations that relate distances, velocities, times, and masses in the 2 different reference frames: that of the train and that of the ground outside.
- E. Einstein called common sense "that layer of prejudices laid down upon the mind prior to the age of 18," and he, at least, was not bound by such prejudices. With a single exception, he showed that all of these common-sense notions are wrong.

III. To investigate how time behaves, imagine 2 spaceships: In one, you are "standing still," while the other contains a fellow experimenter zipping along at a high speed (represented by u) with a special kind of clock.
- A. We designate lengths (d) and times (t) in the moving (to you) spaceship with primes (').
- B. We have your friend allow 1 tick of his light clock.
    1. Velocity equals distance/time, so distance equals velocity multiplied by time.
    2. Since Einstein tells us the speed of light is always constant (which is why we designate it with the letter c), we find $d' = ct'$.
- C. Now you record what 1 tick on her clock looks like to you.
    1. You see the light signal travel a greater distance.
    2. You find $d = ct$.
    3. Meanwhile, the spacecraft has moved to the right a distance ut.
- D. This allows the construction of a simple right triangle with sides ct', ut, and ct.
    1. From the Pythagorean theorem, $(ct')^2 + (ut)^2 = (ct)^2$.
    2. Rearranging terms, it is easy to show that $t' = t(\sqrt{1 - u^2/c^2})$.

3. The astonishing consequence of this little bit of geometry and algebra is that time runs at different rates in the 2 spaceships. When 1 second goes by in our frame, less than 1 second passes in the moving (to us) frame—moving clocks run slow!

E. The square root factor is ubiquitous in relativity, so we should examine its behavior.
   1. When $u \ll c$, the factor is very close to 1—that is, at everyday speeds, which are *much* slower than light speed, we should see no bizarre effects (and we don't).
   2. When u starts approaching the value of c, the factor starts growing; i.e., for $u = 0.9c$, the factor is 0.44, meaning time runs at less than half the speed in the moving frame.
   3. When $u = c$ exactly, the factor is 0; i.e., time stops!
   4. When $u > c$, the factor includes the square root of a negative number, a quantity we call a complex number, hinting that traveling faster than light might present problems.

IV. If the rate at which time passes can be different in different reference frames, what about the other quantities from the Galilean transformations that we also thought we understood?
   A. If time changes, lengths in the direction of motion must also.
      1. The first postulate states that absolute uniform motion cannot be detected; another way of saying this is that there is no one reference frame that is preferred (as the air is in the case of sound).
      2. As a consequence, all observers agree on the relative motion of frames, u; they just don't agree which of them is moving.
      3. Since $u = x/t$ (where x is the distance traveled along the direction of motion) and t changes between frames, x must change also.
      4. To keep u the same for both observers, $x' = x(\sqrt{1 - u^2/c^2})$; i.e., moving lengths contract.
   B. Distance perpendicular to the direction of motion is the one quantity where common sense prevails: $y = y'$, as a proof by contradiction can demonstrate.

- C. With considerably more algebra, one can show that velocities don't simply add and subtract: $v' = (v - u)/(\sqrt{1 + |uv|/c^2})$.
- D. Finally, even masses appear different in the 2 frames: $m' = m/(\sqrt{1 - u^2/c^2})$; i.e., moving masses appear to grow since the square root factor is always less than 1.
- E. It is important to note that the situation here is totally symmetrical since there is no way to determine who is actually moving.
  1. To you (thinking you are stationary), your friend's clock is slow, lengths are shrunken, and masses increase.
  2. To your friend, she is fine in her frame and you are the one suffering all these indignities.
  3. According to relativity, *both* views are correct—there is no preferred reference frame; the quantities really are "relative."

V. Cosmic rays provide an excellent demonstration that these effects are not merely theoretical, nor are they simply apparent changes; these things really happen!
- A. Cosmic rays are the very high energy particles from space we encountered in Lecture Ten producing carbon-14.
  1. They travel very close to the speed of light, so we expect the effects of relativity to be evident.
  2. When they hit the upper atmosphere, they are likely to shatter an atomic nucleus of nitrogen or oxygen and produce a whole shower of subatomic particles.
  3. Among the many types of particles produced are muons (see Lecture Two).
  4. The muons have a half-life of about 1 microsecond (10^{-6} s) before they decay, and they are also traveling very close to the speed of light.
- B. Should we expect to see any muons hitting the ground?
  1. They are created about 30 kilometers up in the outer fringes of the atmosphere.
  2. Even if they were traveling at the speed of light, they should only travel $d = ut$, or $d = 3 \times 10^8$ m/s $\times 10^{-6}$ s = 300 meters, before they decay.
  3. Yet billions of them hit the ground each second.

- C. The effects of relativity are at work.
  1. We see the muon streaking toward us at, say $0.99995c$.
  2. We thus see its clock running 100 times slower, so it takes 10^{-4} s to decay.
  3. In 10^{-4} s, it can travel 100 times farther: 300 m × 100 = 30 kilometers, so it hits the ground, as observed.
- D. How does this situation look from the muon's point of view?
  1. It, of course, thinks (to the extent it could think) that it is only living 10^{-6} s.
  2. However, it sees the distance to the ground as 100 times shorter, since moving lengths contract (remember, it thinks it is stationary and the Earth is rushing toward it).
  3. It agrees, therefore, that it will make it to the ground.
- E. It is always the case that 2 observers in the same place at the same time will agree on the outcome—the muon hits the ground; they just disagree on how it managed to happen.

VI. The concept of simultaneity is also a casualty of the postulates of relativity.
- A. You might see 2 events—flashes to the left and right of you, for example—that you regard as happening at the same time.
- B. An observer moving relative to you will see the flashes occur at different times.
- C. Two observers passing you in opposite directions will even disagree on the order in which the flashes occurred.

VII. The low-velocity, 3-dimensional world we perceive is but a shrunken shadow of the 4-dimensional reality of spacetime.

Suggested Reading:

Abbott, *Flatland*.

Rucker, *The Fourth Dimension*.

Stewart, *Flatterland*.

Questions to Consider:

1. Would you want to go on a journey and come back to Earth in the distant future?
2. What would convince you that clocks in relative motion really do run at different speeds?

Lecture Twenty-One—Transcript
(Almost) Everything Is Relative

Einstein began his development of the theory of relativity with 2 perhaps nonintuitive but straightforward postulates: (1) that absolute, uniform motion cannot be detected; that is, there is no experiment you can perform to establish whether or not you're moving at a constant velocity, and (2) that the speed of light is independent of the speed of its source or the observer. From these 2 postulates, he deduced the logical consequences. This resulted in predictions of bizarre behavior for a body traveling close to the speed of light.

To understand the conflict this engenders with our commonsense notions, we must begin by formalizing the concept of frames of reference. Imagine me standing by the train tracks watching my wife, Jada, go by walking through a train with her little dog, Pixel. Pixel is an absolutely beautiful, highly intelligent, and thoroughly obnoxious little black and tan Pomeranian. So here they are on the train and Pixel's going for a walk while the train is moving by me.

Let's suppose the train is moving to the right with a velocity of 10 meters per second, which we'll designate with the symbol u; u will always be the relative velocity between the 2 frames of reference we're studying. Or, to be more precise, I should say, from my perspective, the train is moving to the right. Because on the train, looking out the window, it appears that it is I who is moving to the left. You may have had this experience before in a train station. Sitting in a train engrossed in your work, waiting for the train to leave, you might glance up out the window and see a train on the next track. It's sometimes very hard to tell whether it's you that's moving forward or that train that's moving backwards.

Pixel is walking in the same direction along the train that the train is moving at, say, 1 meter per second, which is about as fast as his little tiny legs can carry him. Thus, it appears to me that Pixel is moving to the right at 11 meters per second; the 10 meters per second with which the train is carrying him and the 1 meter per second with which his legs are moving.

Let's superimpose on this picture 2 reference frames: one in which I am stationary, the frame of the ground, and one in which Jada is just strolling along slowly, a frame that's attached to the train itself. Since I'm the scientist in this little play, I'll take the role of the observer and will assume that I am truly at rest, ignoring the caveats I mentioned last time about the motion of the Earth around the Sun at 30 kilometers a second, the motion of the Sun around the galaxy at 230 kilometers a second, and the motion of the galaxy through the universe at 600 kilometers a second. But with the respect to the ground, I'm standing still. So we draw the x-axis along the rate of the tracks, the y-axis perpendicular to that, and we'll have the origin right here, where the 2 frames, at this moment, overlap. We'll mark quantities in Jada's frame differently by denoting them with a prime, a little hash: '.

We have carefully synchronized our watches—and they're both very good watches—so at the moment Jada is directly in front of me on the train, we'll mark as time 0 and the 2 coordinate systems will overlap at $t = 0$, $t' = 0$ because our watches match. Because the origins have been arranged to match at this point, we also have that $x = 0$ at the origin and $x' = 0$. They're on top of each other. Likewise, $y = 0$ and $y' = 0$ at our origin; we're on level ground here. And the 2 other things we might want to know about are masses and velocities. The mass of Pixel on the train is 5 kilograms; that's a little over his proper weight, but I must confess he eats a little bit of my good cooking. And that mass is the same for me as it is for Jada on the train; he is just a little chubby to both of us. Finally, there's the velocity with which Pixel is moving through the train, 11 meters per second for me, that's v, and v', the velocity with which Pixel thinks he's moving through the train, is 1 meter per second.

Here's the picture 1 second later, when the reference frame attached to the train has moved with it. Again, you would expect that for me, 1 second has gone by on my watch and 1 second has gone by on Jada's watch. So $t = 1$ second and $t' = 1$ second; x, to me, the location of Pixel in the train, is 11 meters on the x-axis, whereas x' for Pixel is only 1 meter; y is still 0 because the train track is flat, and y' is 0, as well. The mass of Pixel we would expect still to be 5 kilograms, if he hasn't snarfed up anything on the floor, which he's likely to do. He's still 5 kilograms for me, $m = 5$ kilograms, and he's still 5 kilograms in the train, $m' = 5$ kilograms. And finally, since nothing's changed—the train is moving at a constant velocity; Pixel's moving at a constant velocity—velocity, to my point of view, is still 11 meters per second, and v' is still, for Pixel, 1 meter per second.

From these pictures, then, we can derive a set of rules. They're called the Galilean transformations because they were first discussed—not quite in these terms, but the basic ideas were put forth—by Galileo. It's a simple set of equations that relate distances, velocities, times, and masses in 2 different reference frames that are moving with respect to one another; in our case, that of the train and that of the ground outside: x' is going to be equal to $x - ut$, where u is the velocity with which the frame is moving, in our case, the train, and t is the time over which it moves. Clearly, that's true in the case of our circumstance. So, $y' = y$ (that doesn't change), $t' = t$ (we both have watches; they run well; everything works), $m' = m$ (there's no reason to expect the masses to be different), and v', the velocity that Pixel sees in the train, is equal to v, the velocity I see, minus u—minus the velocity with which the frame Pixel is in is moving.

These notions are, after all, just common sense. Here's what Einstein had to say about common sense. He called it "that layer of prejudices laid down upon the mind prior to the age of eighteen." And he, Einstein, at least, was not encumbered by such prejudices. Starting from his 2 postulates, that constant-velocity motion cannot be detected and that all observers agree on the speed of light, he showed that, with 1 exception, all of these commonsense relations of the Galilean transformations are wrong.

Let's start with an investigation of how time behaves. Imagine 2 spaceships. Einstein, in his writing, always used trains, because they were the fastest thing around at the turn of the last century, but we have spaceships, and we can imagine them zipping around at close to the speed of light, so we'll use spaceships in our examples.

In one, you are standing still. Wherever you are, it always feels like you're standing still. Your reference frame is your reference frame. So you're in the spacecraft standing still, while the other spacecraft contains a fellow experimenter zipping along at high speed, again, represented by the velocity u. And you each have a special kind of clock: 2 parallel mirrors between which a light beam bounces back and forth, back and forth, so the clock goes tick-tock, tick-tock, tick-tock.

We again designate quantities in the moving frame—the moving frame to you, that is, in the other spaceship—with the little primes. We have your friend allow 1 tick of her light clock. What happens? The light goes from the top mirror to the bottom mirror.

The velocity, by definition, is equal to distance divided by time, so rearranging the terms, distance equals velocity times time. In this case, the velocity is clearly light, speed c. And since Einstein tells us that the speed of light is always constant—which is, after all, why we designate it with the letter c, for constant—your friend finds that the amount of distance separating her mirrors, d', is equal to c, the velocity, times t', the time on her watch it took the light to get from the top to the bottom.

What do you record when you observe 1 tick on her clock? To you, it looks like this: You see the light travel a greater distance because the light going from the top mirror, once it leaves—these mirrors are both moving along like this—and so when it finally hits the bottom mirror, it bounces over here. It goes this longer distance in your frame d, which is equal to $c \times t$, with t measured in your frame.

Meanwhile, the spacecraft has moved to the right by a distance $u \times t$—u, the spacecraft's velocity, times the same interval of time it took the light to get from one mirror to the other. This allows us to construct a simple right triangle, where the sides are ct', ut, and ct, where ct is the hypotenuse. From the Pythagorean theorem, you will recall that $ct'^2 + ut^2 = ct^2$. That is, the square of the 2 sides is equal to the square of the hypotenuse. Rearranging the terms and taking the square root, it's easy to show that $t' = t$ times the square root of $(1 - u^2/c^2)$. The astonishing consequence of this little bit of geometry and algebra is that time runs at a different rate in the 2 spaceships. When 1 second goes by off your watch in your reference frame, less than 1 second passes in the moving frame, at least the frame that's moving according to us. Moving clocks appear to run slow.

This square root, this factor of the square root of $(1 - u^2/c^2)$ is ubiquitous in relativity, so we should probably take a moment to examine its behavior. When u, the velocity of the reference frame, is much, much less than the speed of light, this factor is clearly very close to 1, because u^2/c^2 is a tiny, tiny number, and 1 minus a tiny number is roughly 1, and the square root of roughly 1 is roughly 1. This is true of what happens at everyday speeds, which are much, much slower than the speed of light. This suggests we should see no bizarre changes in time in our everyday life and, of course, we don't. The square root term is almost always exactly equal to 1 for us, so $t' = t \times 1$, or $t' = t$. Driving at highway speeds, for example, the correction is less than 1 part in 20 million, or less than 1 second in 6 years, and so you're not likely to notice 1 second in 6 years.

When u starts approaching the value for the speed of light, however, the factor starts growing. For example, for $u = 0.9c$, that is, something that's moving by you at a constant speed of $\frac{9}{10}$ the speed of light, this factor, the square root factor, becomes 0.44, meaning that $t' = t \times 0.44$. Time runs at less than half the speed in the moving frame. One second ticks by on your clock and only 0.4 seconds ticks by on the other clock. You have to wait 2.5 seconds before 1 second ticks by on the moving clock.

When u exactly equals c, what happens? Well, u^2/c^2 is 1, 1 − 1 is 0, the square root of 0 is 0, $t' = t \times 0$, $t' = 0$. Time stops. It doesn't matter how long you wait on your watch for time to go by, the time in the moving frame has stopped when the velocity between you and it is equal to the speed of light.

What about when u becomes greater than c? Then u/c or u^2/c^2 is greater than 1, 1 minus a number greater than 1 is a negative number, and this leaves us with the square root of a negative number, something we call a complex number, hinting that traveling faster than light may take us into unfamiliar territory.

We call this effect time dilation. Time gets dilated, or stretched out, in the moving reference frame from our point of view. It's important to note that it's not the clock that is somehow deficient and is, therefore, running slowly. It's time itself that is passing more slowly. A dramatic confirmation of this prediction was carried out by the National Bureau of Standards some decades ago. They took 2 of the successors of Professor Rabi's atomic clocks, I mentioned earlier, left one at their laboratory in Maryland and bought a first-class ticket for the other for an around-the-world flight on a commercial airliner. When the traveling clock returned to Washington and was compared with the clock they left behind, it was slow. Less time had passed for that clock during its travels.

If the rate at which time passes can be different in different reference frames, what about the other quantities from the Galilean transformations that we also thought we understood? If you think about it, a logical analysis of the situation suggests that if time changes, then lengths in the direction of motion must change, as well. Because postulate 1 states that the absolute uniform motion, the constant velocity between the frames, cannot be detected. Another way of saying this is that there is no one reference frame that is preferred, as in the case of sound, there is, in the air. As a consequence, all observers always agree on the relative motions of the 2 frames, the value of u. They just don't agree which of them is moving. To me, Jada and Pixel are moving by on the train. To them, looking out the window, I'm moving by in the opposite direction. We don't agree on that, but we do agree on the size of the velocity, u.

Since the velocity is defined as distance divided by time or, in our picture, $u = x/t$, where x is the distance traveled along the direction of motion, and t changes between the 2 frames, then x must change also for u to be constant. To keep u the same for both observers, x' must equal x times the square root of $(1 - u^2/c^2)$, so the terms in the denominator and the numerator cancel out. That is, a distance parallel to the direction of motion appears smaller in the moving reference frame. We call this length contraction.

What about distances that are perpendicular to the direction of motion? It turns out this is the one quantity where common sense actually prevails: y does equal y', even in the moving frame. And I can show this by a proof called a proof by contradiction, using a meter stick.

Imagine the following scenario: I have a meter stick here and I attach to the top a nice, sharp blade, sticking out like this and another sharp blade sticking down like this at the bottom. And say I equip an adversary over there that's going to run towards me with the same kind of meter stick, with a blade sticking out here and a blade sticking out here. If we charge towards each other—if he charges towards me, say, and I just stand still holding this out—what's going to happen? If y', the moving reference frame, had a contraction in the direction perpendicular to his direction of motion, then when he went by, his meter stick would slice off the top and bottom of mine, because it would be shorter and, therefore, its blades would cut mine off.

Alternatively, if in the perpendicular direction to the direction of motion, to sort of make up for length contraction along the x-axis, the y-axis somehow expanded when he ran by, my ruler would slice off the top and bottom of his, because mine would now be shorter than his. Now, we don't want the rulers to be sliced in either case because that would lead to a preferred reference frame. I could say I am absolutely standing still because my ruler got sliced off. That's not allowed by the first postulate. As a consequence, since we don't allow one reference frame to be preferred over the other, it must be the case that the meter sticks remain the same length in the perpendicular direction and nobody's stick gets the ends sliced off.

With considerably more algebra, one can show that velocities don't simply add and subtract as they do in the Galilean transformation. In fact, it turns out that v', the velocity as observed in the moving frame, is equal to v minus u, as it would be normally, divided by the square root of 1 plus, in this case, $|uv|c^2$, a somewhat more complicated expression. Finally, even the masses appear different in the 2 reference frames; m', it turns out, is equal to m divided by our famous factor, the square root of $(1 - u^2/c^2)$. That is, moving masses appear to grow bigger, since the square root of $(1 - u^2/c^2)$ is always a factor less than 1, and dividing a number by a factor less than 1 makes it larger.

If we adopt Einstein's 2 assumptions about constant velocities and the constancy of the speed of light, we must replace the Galilean transformations with the relativistic ones: $t' = t$ times the square root of $(1 - u^2/c^2)$; $x' = x$ times the square root of $(1 - u^2/c^2)$; y' does equal y, but $m' = m$ divided by the square root of $(1 - u^2/c^2)$; and $v' = (v - u)$ over the square root of $(1 + uv/c^2)$.

It is important to note that the situation here is totally symmetrical, since by postulate, there is no way to determine who is actually moving. To you, who thinks you are stationary and your friend is zipping by in the spaceship, your friend's clock is slow, your friend's lengths in this dimension are contracted, your friend's mass is greater. To your friend, however, who thinks she is happily sitting still and watching you zip by in your spaceship in the opposite direction, she's just fine in her reference frame and you're the one that's suffering all these indignities.

Which, you might well ask, is really the case? Relativity's answer is unequivocal: both. These quantities of mass, and velocity, and length, and time are relative. Indeed, the very notion of absolute space and time itself is an illusion fostered by the fact that we move around in our lives very slowly compared to the speed of light. If we were all given tricycles that went at 50% of light speed at the age of 5, this view—this bizarre, relativistic view—would be common sense to us, because we would have seen these effects all the time.

They would also have the added virtue of being true. But stuck in our slow-moving world as we are, the consequences seem bizarre.

Cosmic rays provide an excellent demonstration that these effects are not merely theoretical, nor are they simply apparent changes; these things really, really happen. Cosmic rays are these very high-energy particles produced in the shock waves from exploded stars in distant regions of interstellar space, which we encountered in previous lectures. They produce the carbon-14 in our atmosphere, they produce the thermoluminescence in roof tiles, and other terrestrial phenomena. Unlike most things we encounter, cosmic rays actually do travel very close to the speed of light. With a velocity approximately equal to c, we should expect to see some of these weird effects of relativity and that they'll become important. Indeed, they do.

When a cosmic ray hits the upper atmosphere, it's very likely to shatter an atomic nucleus of nitrogen or oxygen or some other atmospheric gas and produce a whole shower of subatomic particles. Among the many types of particles produced are muons. Some of you will remember from Lecture Two that muons are a heavy version of an electron, a heavy lepton. Now, muons aren't around—you don't find them under your kitchen sink—because they're very unstable. They have a half-life of 1 microsecond—10^{-6} seconds—before they decay, so they don't exist in our everyday world. But they're produced constantly, all the time, by collisions of cosmic rays with atmospheric atoms. The question is: Should we expect to ever see any of these muons hitting the ground? They're created about 30 kilometers up, in the outer fringes of the atmosphere. So here's the surface of the Earth and here's the cosmic ray coming in, slamming into an atom and producing a whole shower of muons.

Since muons have mass, they cannot travel at the speed of light, but they come very close to the speed of light. So for an approximation, let's just assume for the moment that they do. They will travel a distance, d, which is equal to c, the velocity they're moving, times t, the time they have to live. Now, the velocity with which they're moving is just a smidgen less than 3×10^5 kilometers per second, the 300,000 kilometers a second of light speed. And the time they have to live is 10^{-6} seconds. And so we should expect they can only go 0.3 kilometers before they decay. Yet billions of them hit the ground each second. How can they do that? The effects of relativity are at work. Here we see the muon streaking toward us, from its creation in the upper atmosphere, at 99.995% the speed of light. This is, in fact, a typical speed for a muon. Using the time dilation equation, we can then calculate how slowly, from our perspective, that rapidly speeding muon will be decaying. In our frame, for the muon, t' will be equal to t on our watches times the square root $(1 - u^2/c^2)$. And plugging in the numbers, t as 10^{-6} seconds and u as 99.995% the speed of light, we find that $t' = 10^{-4}$ seconds. Its clock is running 100 times slower than it would if it were sitting on our bench in the lab. To us, it takes 10^{-4} seconds to decay.

How far can it go in 10^{-4} seconds? Clearly, it could travel 100 times farther than it would if it were dying in 1 microsecond. That's 0.3 kilometers, the distance it would have traveled in 1 microsecond, times 100, or 30 kilometers. Guess what? That's where it was created, 30 kilometers up. It hits the ground before it decays, as we observe it.

How does this situation look from the point of view of the muon? It, of course, thinks—to the extent that a muon could think—that it's only living 10^{-6} seconds, because it's living in its frame where it's standing still. However, it sees the distance to the ground as 100 times shorter, since it sees moving lengths contracted and it sees the Earth rushing towards it at an enormous speed. It agrees, therefore, that it will make it to the ground, because it only has 300 meters to travel before the Earth hits it and it's got 10^{-6} seconds to make that trip. Bingo, it hits the ground. It is always the case—and this is absolutely critical to understand—that 2 observers in the same place at the same time will agree on the outcome. The muon hits the ground. They just disagree on how this managed to happen.

The concept of relativity of space and time, the postulates themselves, are all relative. Even simultaneity, whether 2 events occur before or after each other, is relative. Picture 2 rocket ships traveling in opposite directions at, say, 80% the speed of light, like this. The 2 pass a third, stationary ship, the one on which you're located, sitting still in space. Stationary, of course, because it's yours. Now, just when the 2 spacecrafts pass, 1 above you and 1 below you, you push a button that triggers a camera on either end of your ship to take a flash picture of the other spaceships.

In this case, ship A, traveling to the right, sees the following sequence of events: Flash Y occurs first and flash X occurs later. Ship B, traveling in the opposite direction, sees flash X occur first and then Y. Actually, from B's point of view, what he actually sees is the following: X flashes when the tail of A is aligned with the nosecone of B, and Y flashes when the nose of A is aligned with the tail of B. B also sees a tiny, shrunken ship A because it's moving by him so rapidly and its length is contracted and concludes with confidence that X happened before Y. Ship A, of course, sees exactly the reverse: a tiny ship B, flash Y happening before flash X. You, on the other hand, see both ships as shrunken in size—one's moving this way and one's moving this way, but at the same speed, their length is contracted by the same amount—and the flashes, X and Y, you see as simultaneous. But there is no way that you will convince your colleagues that the 2 flashes actually happened at the same time. After all, they both saw the order of events in their particular reference frames.

The painting behind me is a 2-dimensional representation of a forest at night. It's a flat piece of paper: x, y, that's it. Yet my brain perceives, when I look at it, a 3-dimensional scene: Some of the trees are in front, some of the trees are behind and block our view. What happens when we view a truly 4-dimensional world of spacetime through 3-dimensional eyes?

A useful guide to this dilemma is provided by something like this stack of paper. In the book *Flatland*, a British math teacher in the late 19th century combines simultaneously a marvelous searing satire of the Victorian view of women and a look into higher dimensions. In *Flatland*, the women are straight lines. It's a 2-dimensional world. The men are triangles—straight lines, of course, meaning that women are 1-dimensional—but the men are triangles, and the higher one goes in the political hierarchy, the more sides one acquires. So a square is ahead of a triangle and an octagon is ahead of a square.

One day, our hero and his wife, the triangle and a straight line, are sitting in their little 2-dimensional square house and a sphere comes to visit. Now, what would it look like if you were sitting in your 2-dimensional living room and sphere came to visit out of another dimension? They're sitting there and a dot appears in the middle of the room. It then slowly grows to a bigger and bigger circle as the sphere passes through the 2-dimensional world of Flatland. And then it reaches a maximum size and starts shrinking again and disappears. It doesn't come in the door, it doesn't come in the windows, it comes out of nowhere from another dimension.

In a sense, our position is the same as the Flatlanders'. In reality, we inhabit a 4-dimensional spacetime, but we observe it with 3-dimensional sensibilities. If I hold up this stack of paper in a bright light and only observe its shadow, I see a 2-dimensional image which, in this position, tells me the length and height of the shape. But if I rotate the book through time, its shadow changes. What is the true book like? In our slow-speed, 3-dimensional world, we observe only the shrunken shadows of objects in the reality of 4-dimensional space. This is the lesson of relativity.

As the muon makes clear, relativity works, it predicts the outcomes of real experiments in the real world. We are spared its more bizarre consequences because we travel around at such slow speeds. But when we want to explore the expanding universe to find the origin of the particles of which it is composed, relativity is critical. In particular, we must look at 2 of the most famous of relativity's predictions: first, that nothing travels faster than the speed of light and, second, that matter can simply disappear.

Lecture Twenty-Two
Matter Vanishes; Light Speed Is Breached?

Scope:

Relativity does not forbid faster-than-light travel. However, it does impose severe consequences for such an occurrence. In particular, using a space-time diagram, it is straightforward to show that in a world with hyper-rocks (theoretical particles that travel faster than the speed of light), it is possible for a window to break before the child that threw the rock at the window was born. This is not forbidden by physics, but it is profoundly disturbing and as yet unwitnessed in our surveys of the universe.

Outline

I. Let's begin with the challenge I offered a couple of lectures ago: Define time.

   A. Not only does subjective time flow at different rates; we have also seen, as relativity tells us, that objective time runs at rates dependent on one's velocity.

   B. My old cosmology professor, Ted Harrison, had a good metaphor for time: a wave of vividity.

II. If even time and space are not absolutes any more, is anything?

   A. The Greeks recognized that one could calculate the distance between 2 points on a 2-dimensional plane by using the Pythagorean theorem.

   B. Descartes generalized this to 3 dimensions. The equation $d^2 = x^2 + y^2 + z^2$ gives the distance, or space interval, between 2 points in 3-dimensional space.

   C. Einstein generalized space and time into a 4-dimensional spacetime.

   1. One might expect a spacetime interval D to be given by $D^2 = (\text{space})^2 + (\text{time})^2$, but as usual, one's common sense is incorrect.

   2. In fact, $D^2 = |(\text{time})^2 - (\text{space})^2|$ where the minus sign is an indication of the bizarre properties spacetime has.

- **D.** The spacetime interval D so defined is invariant—all observers always agree on the this interval between 2 points in spacetime.
  1. The shortest distance between 2 points is not a straight line.
  2. Observers disagree on lengths in space and intervals of time.
  3. Observers disagree on whether or not 2 events are simultaneous.
  4. But everyone agrees on D.
- **III.** To represent spacetime, we simply draw a space-time diagram with space on one axis (usually the x-axis) and time on the other.
  - **A.** To make the interpretation easy, we label the time axis in seconds and the space axis in light-seconds (or years and light-years when talking about the universe).
  - **B.** An object sitting still is represented by a straight vertical line on a space-time diagram called a world line.
  - **C.** A object traveling from one point to another at a constant speed looks like a tilted world line; an accelerating object is a curved line.
  - **D.** Light has a special place in space-time diagrams.
    1. With the axes labeled conventionally, since light travels 1 light-second in 1 second, light is represented by a 45° line.
    2. Drawing all 4 diagonal lines from a given location forms the light cone.
    3. The light cone divides the diagram into a time-like part, containing an absolute past and future, and a space-like part, which is at present unreachable by the observer and contains a relative past and future. A star exploding 10 light-years away cannot be discovered until 10 years has passed.
  - **E.** The invariant spacetime interval D reveals interesting aspects of the diagram (and of reality).
    1. The longest interval between points A and B is the straight line connecting them.
    2. A shorter interval would be obtained by traveling away from the location and then returning.

3. This property of spacetime gives rise to the famous twin paradox.
    4. The special role of light is apparent: The spacetime interval experienced by a photon of light is precisely 0—light does not experience spacetime.
  F. Different observers will draw different space-time diagrams.
    1. The lines of one observer will be tilted with respect to those of the other, since each one thinks he or she is stationary.
    2. However, most importantly, the light cones *do not* tilt, since the speed of light is constant for all observers.

IV. Space-time diagrams make it easy to illustrate the consequences of faster-than-light travel.
  A. Consider a scene in which you are standing in a room looking out the window and see a nasty little kid pick up a rock and hurl it at the window.
  B. You see the sequence of events as (1) kid is born, (2) kid picks up the rock, (3) kid throws the rock, (4) rock hits the window, and (5) the window beaks—only common sense.
  C. Now imagine the kid has a laser gun in stead of a rock.
    1. The space-time diagram shows the laser light traveling along a light cone.
    2. You see the kid born, and then the gun fires and the window breaks simultaneously.
  D. Now imagine that the kid has a hyper-rock that can travel faster than the speed of light: The sequence of events you witness is clear: (1) the window breaks, (2) the kid is born, and (3) he throws the rock.
  E. The second situation is not forbidden by the laws of relativity.
    1. My late Columbia colleague Gary Feinberg proposed the "tachyon," a particle that goes faster than light, and worked out its properties.
    2. Tachyons only appear in science fiction—not, to date at least, in the real world.
    3. The only price for admitting them to one's world is to abandon cause and effect, all ye who enter here.

V. Antimatter, however, does exist in the real world.
  A. It generally does not last for long.
    1. If an antimatter particle encounters a matter particle of the same type, they annihilate completely, leaving 0 mass behind.
    2. This does not violate the law of conservation of mass when it is extended by relativity to be a conservation of mass-energy.
    3. The conversion is governed by the relation $E = mc^2$.
  B. Thus mass can turn into energy and energy back into mass, a process that occurred readily in the early universe and is critical to our existence today.

Suggested Reading:

Fraser, *Antimatter*.

Questions to Consider:

1. Do you believe we will eventually overcome the speed limit of light?
2. Antimatter-matter annihilation is an incredibly efficient source of energy. Why do you suppose we are not working to exploit it?

Lecture Twenty-Two—Transcript
Matter Vanishes; Light Speed Is Breached?

Science fiction writers, 10-year-old boys, and many of the innumerable crackpots who send me their theories of the universe each year are united by 1 conviction: that things *can* travel faster than the speed of light. In fact, relativity does not impose c as an absolute speed limit. But it does spell out the consequences if that limit is exceeded. After this lecture, it will be up to you to decide if you're willing to live with those consequences.

Science fiction writers and 10-year-old boys also love antimatter. But that's just fine, because antimatter really exists. Indeed, as I explained way back in Lecture Two, every particle in the putatively fundamental list of particles, the 6 quarks—up, down, strange, charm, top, bottom—and the 6 leptons—muon, tau, electron, and their relevant neutrinos—each has an antimatter mate This is not just theory; we can make and we can detect each of these antimatter particles—antimuons, antiprotons, antineutrons, antineutrinos. The antielectron, or positron as we've called it, is the instigator of half of all radioactive beta decays. The positron's emergence from the nucleus signals the transformation of a proton into a neutron and the demotion of that element 1 step in the periodic table. Antimatter is frequently produced in the cosmic ray collisions with the air we've discussed. And as we shall see, antimatter played an absolutely crucial role in the first moments of the universe when all matter was born.

Relativity theory is the theory that allows us to explain both the consequences of faster-than-light travel and the mechanism by which antimatter arises. We'll begin with the former since it is important for deepening our understanding of space and time. And we'll end with the latter because it plays such a central role in the early days of the universe, when the building blocks of our atomic historians were created.

Let's start with the challenge I offered a couple of lectures ago: Define time. Not only does subjective time that we experience in our lives flow at different rates; we have seen that relativity tells us that even objective time kept by accurate clock runs at rates dependent on one's velocity. My old cosmology professor, Ted Harrison, had a good metaphor for time, although I'm not sure I endow it with the definition. He called time a wave of vividity, and he drew it like this.

At the moment we are now, our vividness is at a peak. I vividly feel the lights on my face, the pencil in my hand, the paper under my other hand. The moment is vivid. A minute ago, things were almost equally vivid; I was standing here talking into the same cameras, with the same lights, and the same paper, and the same pencil in my hand. Yesterday, well, that's a little less vivid; I don't remember exactly what I was doing at this time yesterday. A year ago, I certainly don't remember precisely what I was doing, so the vividity index falls off and generally slopes away to close to 0. I certainly don't remember what I was doing when I was 2 years old. There are little peaks occasionally—for example, it's still very vivid to me the time when the zipper on my jacket got stuck in first grade and the teacher had to rip it off over my head, including taking with it my shirt, such that I was standing half naked in front of my first-grade compatriots. So there are vivid moments like that. The wave of vividity also extends a little bit into the future sometimes. If I step off the curb in New York without looking and turn to the left and see a bus only 2 meters away, it's very vivid to me what's about to happen. Still, this is a nice analogy for time, but it's not a definition of time.

We agree that time and space are not absolutes any more once we adopt the theory of relativity. Is there anything that all observers can agree on? They can't agree on time; it runs at different rates. They can't agree on simultaneity. They can't agree on distances. They can't agree on masses. Is there something that's absolute and enduring for all observers?

The Greeks recognized that one could calculate the distance between 2 points on a 2-dimensional plane using the Pythagorean theorem. You have a coordinate system with an origin. You have an x value and a y value, you plot the point at that location, and you draw a line. If you take $x^2 + y^2$, it's equal to the distance, d^2, between the 2 points. The 17th-century mathematician Rene Descartes generalized this to a 3-dimensional world, the one in which we live. If you draw a 3-dimensional graph on the piece of paper and plot a point at some position x, y, and z, in or out of the page, then the distance from the origin to that point, d^2, is simply given by $x^2 + y^2 + z^2$.

Einstein, now, has generalized space and time into a 4-dimensional spacetime. One might expect, in this case, that the spacetime interval, D, will be given by $D^2 = x^2 + y^2 + z^2 + t^2$, but as usual in relativity, one's commonsense notions are incorrect. In fact, in relativity, the spacetime interval, the distance between 2 points in the 4-dimensional spacetime, $D^2 = |(\text{time})^2 - (\text{space})^2|$, where "space" here is meant to stand for the 3 normal dimensions, x, y, and z, and where the minus sign is an indication of the bizarre properties that spacetime has.

This spacetime interval, D, defined in this way, is, in fact, invariant. All observers will always agree on this interval between 2 points in spacetime. They disagree, as I've said, on length in space and intervals of time. They disagree on whether or not 2 events are simultaneous. But everyone agrees on D. It will turn out that the shortest distance, the shortest spacetime interval between 2 points, is not a straight line. Yet another bizarreness of relativity.

To represent spacetime, we simply draw what we call a space-time diagram, with 1 dimension of space, usually that on the x-axis, and 1 dimension of time, which we draw on the y-axis. In reality we would try to represent the full 4 dimensions, but I have trouble even drawing 3-dimensional figures on a 2-dimensional piece of paper, so let's stick to the 1 x dimension for now as representing the dimension through space and the y-axis being the dimension through time. To make this easy to interpret, we label the axes such that the time axis, say, is labeled in seconds, and the x-axis is labeled in light-seconds, that is, the distance light travels in 1 second. Or, if we're talking about events out in the universe, we might label the y-axis in years and the x-axis in light-years, the distance light travels in a year.

This diagram is made up of many, many events. A point in this diagram represents an event because it is something that happens at some particular place in space at some particular moment in time. The place in space is located on the x-axis, the place in time is located up the y-axis, and the point there represents the event.

An object sitting still in a space-time diagram is represented by a straight vertical line, because time is certainly passing for that object, but it's not moving anywhere in space. We call this a world line. An object traveling from one point to another at a constant speed in the space-time diagram looks like a tilted world line, because it is moving through space—so along the x-axis—as time passes. So it makes it a tilted line. Accelerating objects are a curved line. They start off standing still—so a vertical line—and then they gradually tilt away more, and more, and more as they cover more space in a given amount of time.

Light has a very special place in the space-time diagram. With the axes labeled conventionally—since light travels 1 light-second in 1 second, 2 light-seconds in 2 seconds—light will be represented by lines at 45° covering 1 light-second in 1 second, 2 light-seconds in 2 seconds.

Drawing the 2 diagonal lines from a given location forms what we call the light cone, because it looks sort of like a cone. The light cone divides the diagram into a time-like part, containing an absolute past and an absolute future, and a space-like part, which is at present unreachable by the observer at the point where the lines cross and contains relative past and future. For example, if you're sitting here, at the origin of these two 45° lines, anything in the backwards part of your light cone, you've already seen, because light from any point, any event—an explosion, a birth, a death—in that back light cone, the light will have had time to reach to you and you will already know about it. The things in the forward part of the light cone, you absolutely have not got to yet; an event happening there will not come into your ken until you reach that point in the diagram.

But an event that happens outside of the light cone is neither in your past or your future, or rather, it could be both in your past and in your future. It could be in your past in the sense that if you drew a line horizontal across the diagram from where you are today, you might think everything below that is in the past. But as I discussed in talking about Betelgeuse, the star in Orion that's about to explode, if the star is 400 light-years away and it exploded 300 years ago, we won't know about it yet, because it will lie outside of our backwards light cone. Only as we move forward in time and our light cone encompasses more, and more, and more of the space-time diagram will this supernova eventually come into view.

The invariant spacetime intervals D reveal interesting aspects of the diagram and of reality. The longest spacetime interval between points A and B on a stationary world line, that is, a vertical line, is that distance, the straight line that connects points A and B, because D^2 is time squared minus space squared, and the space motion in that case is 0. You're in the same place. You're in the same point on the x-axis. So $t^2 - 0$ is the maximum amount that you can get. A shorter interval would be obtained by traveling away from the location and then returning again, because the same amount of time has gone by, but you get to subtract off the space through which you've traveled. We'll see how this works in a moment.

We actually saw this effect when discussing the atmospheric muons last time. Let's see what that situation looks like on a space-time diagram. From our perspective, the muon has to travel a long way through space, from the top of the atmosphere, over here on the x-axis, to the Earth, over here on the x-axis. Now, it's traveling very close to the speed of light, so it's almost at a 45° angle. What kind of spacetime interval does it go through? It goes through that amount of space and that amount of time, $D^2 = t^2 - x^2$. And you see, it's a very small spacetime interval which it has gone through. It looks like a long time to us to get there—that's its moving clock running 100 times slower than normal—but in spacetime it's a very short distance. From the muon's perspective, it sees the Earth rushing upon it as soon as it's created and, a tiny interval later, slamming into it. It lives what it thinks is a normal amount of time, a small amount on the time axis, but that's okay, because it doesn't have very far to go. The space-time diagram shows how the unusual situations, the non-commonsensical situations of relativity, are easily explained.

The property of spacetime represented by this effect gives rise to the famous twin paradox. Suppose you have a twin who becomes an astronomer, an adventuresome type who decides to go off on an interstellar journey. You, well, you'd rather stay at home and watch Teaching Company DVDs. Here's you on the diagram, staying at home in the same place as Earth, as a vertical line. And here's your twin that goes accelerating away and then cruising at nearly the speed of light, off, perhaps, to see whether Betelgeuse is really going to explode soon. He gets there, takes his measurements, decides that an explosion is imminent, and promptly turns around to return and alert the astronomers on Earth so they'll be ready for this monumental event. Tragically, when he arrives back home, he finds 2 things have happened: First, everyone already knows the star has exploded, and second, his twin is dead. Let's analyze this scenario on the space-time diagram.

Because we're talking about traveling hundreds of light-years through space, we'll label the x-axis in light-years and the y-axis in years. Here's you, your vertical world line indicating that you're sitting still on the Earth. And here's the path of your twin's trip, out to the vicinity of Betelgeuse and then back again. What happened? After he started his return, the star exploded, and while he was traveling fast, the light from the explosion traveled precisely at the speed of light—spreading out in all directions and, in particular, moving towards Earth at light speed, faster than his rocket could travel. Obviously, then, it arrived before he did, so the explosion was old news by the time he returned. In addition, we must calculate the spacetime intervals represented by your journey through spacetime and his. Your journey, sitting there watching the DVDs, is given by $D^2 = t^2 - x^2$, and since the space you covered was limited to that between your couch and the refrigerator, it's essentially t^2, because the space squared is 0, especially on this diagram, where the x-axis is labeled in light-years, 1 light-year being 6 trillion miles. In this scenario, 801 years have gone by since your twin left on his journey. It's very likely, I'm afraid, that your personal world line will end before 801 years is up.

Your twin, however, will have covered distance $D^2 = t^2 - x^2$, but in his case, x^2 is a very big number; he's gone 400 light-years out and 400 light-years back. If, for example, he were traveling at 99.9% the speed of light throughout most of his journey, the spacetime interval through which he has traveled, going from the time he left to the time he returns at Earth, is only 36 years. He'll be a little older, but just fine—there to meet your great-great-great-great-great-great-great-great grandchildren.

But wait, you should be saying: I thought you said last time that the situation was symmetrical. I see his clock running slow because he's moving, but he should see my clock running slow, too, so why aren't I here to greet him? Well, remember; remember the conditions of Einstein's postulates. They hold in situations where velocities are constant. Clearly, if you accelerate from Earth to nearly the speed of light, go out to Betelgeuse, turn around, and come back, your velocity isn't constant. I've slipped a new condition into this story; accelerations are involved.

These accelerations and decelerations when he turns around at Betelgeuse violate the strict conditions in which special relativity applies. General relativity, which took Einstein another 11 years to figure out, takes care of this. So the situation I've described would occur the way it does, but the simple equations of the relativistic transformations I've taught you don't really take that into account.

The special role of light is apparent on a space-time diagram. The spacetime interval experienced by a photon, D^2, moving from here to here on the diagram, is 0, because it's the time squared minus the space squared. And since, by definition, this is a 45° line, those 2 distances are equal, $t^2 - x^2 = 0$. Light does not experience spacetime. The photons we will meet, which come to us from the very edge of space and time, haven't gotten tired along their 13.7-billion-year journey—no time has elapsed for them at all.

Different observers draw different space-time diagrams because each observer believes he or she is sitting still and is, therefore, represented by a vertical line. The lines of one observer traveling with respect to the other will be tilted with respect to each other—you just take the whole graph and rotate it, since each observer thinks he or she is stationary. However, most importantly, the light cones on these tilted space-time diagrams do not tilt because the speed of light, according to Einstein's second postulate, is constant for all observers.

Space-time diagrams make it easy to illustrate the consequences of faster-than-light travel, the first of the 2 keys of relativity I want to talk about today. Consider a scene in which you are standing in a room looking out the window and see a nasty little kid pick up a rock and hurl it at the window. So, space-time diagram: We draw a t on the y-axis; space, x, on the x-axis. We draw a vertical line for your world line, a vertical line near you representing the window, and a vertical line farther away representing the kid who picks up the rock. The sequence of events that you see is clear. The kid first picks up the rock, then you see the kid throw the rock—these are all represented by light cones on the diagram as time moves forward—then you see the rock hit the window, and then you see the window break. It's only common sense.

Imagine instead this is a high-tech kid that has a laser gun instead of a low-tech rock. His aim is the same, to break your window. Let's plot what you would see in a space-time diagram in this scenario. The kid is born, down here sometime, he grows up, he enters your past light cone so you know of his birth. The kid picks up his laser gun and he fires the laser gun. The laser, of course, is just light, so it travels along the light cone. From your perspective, it (a) leaves the gun, (b) hits the window, and (c) you see it simultaneously because it travels at the speed of light and you can't see anything that happens before the light manages to reach you.

This does not mean that light takes no time to travel this distance through space. Light has a finite velocity, 300,000 kilometers per second. And so, velocity equals distance over time, that means time equals distance over velocity, $d/300,000$ kilometers per second; that's a real number. It takes a finite amount of time for the light to get from the laser gun to your window. However, it does mean that the light does not experience spacetime—its spacetime interval is 0—and it means you do not see a gap between the time when the gun is fired, the window breaks, and that information reaches your eyes. The gun really did fire before the laser beam got to the window, but you cannot observe that because you can only observe the light.

Here's the challenge for you faster-than-light people: Consider that the kid, rather than a rock or a laser, has a hyper-rock that can travel faster than the speed of light. Again, we draw the space-time diagram. Time on the y-axis, space on the x-axis, the vertical line for you, the vertical line for the window out which you are observing this scene. The sequence of events, if we draw the kid and the hyper-rock in this way, comes to you as follows: The window breaks, the kid is born, and then he throws the rock.

This situation is not forbidden by the laws of relativity. Indeed, my late colleague Gary Feinberg invented tachyons, theoretical particles that behave like hyper-rocks, that go faster than the speed of light. And Gary worked out all their properties. They're a little bit strange; they travel backwards in time, they have negative energy, etc. Gary was a connoisseur of science fiction and would not be unhappy to see today that his tachyons have gained life in popular literature, although he certainly didn't expect that their presence would ever be discovered in the real universe. Because traveling faster than light eliminates cause and effect. It is not forbidden by relativity; relativity has nothing to say on the subject except to allow one to compute the properties of entities that travel faster than the speed of light. The price for allowing it, however, is "abandon cause and effect, all ye who enter here."

As a brief aside, I should note that it is the speed of light in the vacuum of space that no object with mass can match. Light actually travels more slowly through matter—through air, through glass, through water—and it is possible for particles in air to exceed the speed of light *in air*, although not the 300,000-kilometer-per-second limit. Indeed, we actually observe particles traveling through air faster than light travels through air. When they do, they emit a ghostly little blue light, which is called Cerenkov radiation, and we've actually detected this signal in our atmosphere with telescopes, where the signal is produced by high-energy cosmic rays that we keep discussing that slam into the atmosphere and make the radioactive isotopes and muons that we've discussed. Travel faster than light in a vacuum is not happening.

Antimatter is another matter, so to speak. One of its most bizarre properties is that it disappears when it encounters normal matter. I mean, disappears completely—its mass no longer exists. This would seem to violate the law of conservation of mass, that pillar of 19th-century science that was so central in the development of both chemistry and of physics. And, indeed, it does violate that law. You start out with a proton and an antiproton. Both of them are perfectly well-defined objects. They both have mass. You can put them on a scale. They weigh 1.67×10^{-24} grams—yes, it's a small number, but that's okay. They have mass; you can record it. You bring them together, you end up with 0, $m = 0$, precisely.

Again, relativity comes to the rescue to explain this circumstance. Yet another consequence of Einstein's 2 postulates and the redefinition of time and distance they require is that mass is not constant either. As we saw last time, in the relativistic transformation, mass is one of those relative properties which depends on the speed of the observer: m' equals m divided by the famous square root of $(1 - u^2/c^2)$. Pixel on the train looks even fatter than he does in reality to me standing on the ground. This notion, alone, already does violence to the idea of conservation of mass. But the full implications of the theory of relativity go well beyond this. By following the logical consequences of relative times, distances, velocities, and masses, and developing additional such concepts, like relativistic momentum and rest-mass energy, but without involving any terribly complicated mathematics, one reaches a remarkable conclusion: that mass and energy are equivalent quantities, that you can convert one to the other as long as the total of the 2 is constant.

The famous expression of this equivalence, perhaps the most famous equation of all time, is $E = mc^2$. E is the total energy, m is the mass, and the proportionality constant between them is c—unsurprisingly, that speed of light again, which keeps showing up everywhere—the speed of light squared. If one has a photon, the ultimate expression of pure energy, absolutely 0 mass, it can spontaneously, if it has enough of that energy, turn itself into a particle/antiparticle pair, both members of which *do* have mass—you can catch them, you can put them on a scale, and you can weigh them. Symmetrically, any antiparticle/particle pair that meets up will quickly annihilate, their masses disappearing completely into the energy of a photon that speeds away, obviously, exactly at the speed of light.

Einstein's relativity provides yet another example of how, on scales with which we cannot be intimately familiar, in realms far from the direct experience of our senses, things can be very different than they are in our prosaic world of human scales. Having yet to discover tachyons, we must rely on photons, traveling merely *at* the speed of light, to bring us information about the most distant reaches of space and, therefore, of time. In particular, we are now prepared to examine the oldest photons in the universe, the afterglow of matter-antimatter annihilation in the big bang itself, in order that we may complete our history of the atoms and their constituents.

Lecture Twenty-Three
The Limits of Vision—13.7 Billion Years Ago

Scope:

The oldest stars we can see are nearly 13 billion years old. As their light travels to us through space, it is modified by the atoms it encounters along the way. In particular, the cold clouds of gas that have yet to condense to form galaxies reveal the ratio of normal hydrogen to heavy hydrogen (deuterium). Since both isotopes were produced in the first 3 minutes after the big bang, this ratio provides us with a direct measure of the temperature and density of the universe at the moment of its birth. Using telescopes sensitive to microwaves, we can peer further into the distant past to a point just 380,000 years after the universe began, when the very first atoms formed.

Outline

I. The spacetime of the universe has been expanding for 13.73 billion years.
   A. In the first 2 decades of the last century when Vesto Slipher was, like Douglas of tree-ring fame, a young astronomer at the Lowell Observatory, he undertook a program to measure the velocities of the "spiral nebulae."
      1. These nebulae were then thought to be solar systems in formation within our Milky Way.
      2. He measured their velocities with respect to Earth by looking for wavelength shifts in their light.
      3. He found the astounding result that 17 out of 21 had their light shifted to the red end of the spectrum and were thus inferred to be moving away from Earth.
   B. The cause of this red shift is the Doppler effect.
      1. When an object emitting a wave moves toward an observer, the wave crests are bunched together, giving the appearance of shorter wavelengths.
      2. Conversely, when an object moves away, the wave crests appear farther apart.
      3. With sound, this leads to the familiar ambulance effect.

4. With light, since color is a manifestation of wavelength, it leads to a blue shift for objects moving toward the observer and a red shift for objects moving away.
C. At the same time, Edwin Hubble discovered that he could see individual stars in the spiral nebulae.
1. This implied that they might be separate galaxies from our own.
2. By using a class of regularly pulsating stars, he derived rough distances to these nebulae and found them to lie far outside the Milky Way.
3. Incorporating Slipher's data, he derived the astonishing result that the farther away a nebula is, the faster it is moving away from us.
D. The naive interpretation of Hubble's observation is very disturbing.
1. If all galaxies are running away from us, it would seem we must be at the center of the universe.
2. Furthermore, if the farther away they are, the faster they are fleeing, then they must all know where we are.
3. Both thoughts violate the cosmological principle: We do not occupy a special place in the universe.
4. As a human-conceived principle, the cosmological principle could, of course, be incorrect.
5. Nonetheless, it is worth thinking about alternative explanations for Hubble's extraordinary discovery.
E. The simplest alternative is that spacetime is not static but expanding.
1. If the spacetime is stretching, the distances between all pairs of galaxies will increase.
2. Furthermore, the greater the original distance between 2 objects, the faster that distance grows.
3. Galaxies arrayed on the surface of an inflating balloon provide a useful visual analogy.
F. Numerous independent lines of evidence confirm that the universe is indeed expanding from a much denser phase in the distant past; this framework for interpreting our observations is called the big bang model.

II. The red shift induced by expansion provides a handy way to measure the distances of remote objects.
  A. Since all hydrogen atoms in the universe are identical, their electron jumps lead to well-known emitted wavelengths.
  B. Measuring the observed wavelength is straightforward.
  C. The ratio of observed to emitted wavelengths gives the factor by which the universe has expanded since light left the emitting galaxy.
    1. As the waves travel through the expanding spacetime, their crests grow farther apart.
    2. They stretch by a factor of 2 when the spacetime has doubled in size.
  D. Having measured the rate of expansion by using objects of known distance, the red shift thus directly gives an object's distance.
  E. The most distant galaxies we have detected are at a red-shift factor of 6.5.
    1. The universe was about 7 times smaller.
    2. The universe was $7 \times 7 \times 7 = 350$ times denser since, as we shall see, the number of fundamental particles has not changed since that time.
    3. The light left these galaxies about 12.9 billion years ago, when the universe was only 850 million years old.
  F. How can we use this ancient light to learn more about the history of atoms?

III. A few minutes after the big bang, only 6 isotopes were present in the universe, 2 each of hydrogen, helium, and lithium: ^{1}H, ^{2}H, ^{3}He, ^{4}He, ^{6}Li, and ^{7}Li.
  A. The relative amounts of each of these isotopes provides a strong constraint on the earliest moments of the universe.
  B. Measuring their abundances from nearly 13 billion light-years away is a challenging task.
    1. At this time, most of the matter had not yet collapsed into galaxies and was spread throughout space in tenuous clouds.

2. The initially densest regions, however, had collapsed, and some had produced giant black holes a billion times the mass of the Sun, which gobbled up surrounding material at a prodigious rate.
3. These feeding black holes generated more than 1000 times the light of a galaxy of 100 billion stars and can easily be seen across the vast distances involved.
4. As their light passed through the clouds of gas along the way, deuterium (^{2}H) atoms absorbed the light at the precise wavelengths corresponding to their electron jumps, which differ slightly from those of normal hydrogen (^{1}H).
5. Detecting both ^{1}H and ^{2}H in the same cloud allows us to calculate their abundance ratio; since this gas is unpolluted by the effects of nucleosynthesis in stars, it represents the value for the primordial material formed in the big bang: 30 deuterium atoms for every 1 million hydrogen atoms.

C. The other 4 isotopes are difficult to measure at high red shift, so their determinations are made more locally.
1. Galaxies yet to experience much star formation provide relatively pristine samples of primordial material.
2. The very oldest stars in our galaxy likewise are made of material close to that of the original composition.
3. From such observations, the primordial abundances appear to be as follows: 22% ^{4}He; 50 parts per million ^{3}He; 3 parts per billion ^{7}Li; and Li-6 is as yet undetected, so less than 1 part per trillion.

D. These abundances are key to constraining the conditions in the first 3 minutes of the universe.

IV. The oldest photons in the universe give us a direct picture of the universe's structure and conditions when it was only 380,000 years old.

A. The most distant light-emitting objects we have discovered lie roughly 12.9 billion light-years away.
1. When that light was emitted, the universe was roughly 850 million years old.
2. It may well have taken that long for diffuse gas to collapse due to gravity and to form objects—stars and black holes—bright enough for us to see.

B. However, observing the sky at microwave wavelengths (1 mm–1 cm), one sees a uniform, diffuse glow called the cosmic microwave background (CMB).
  1. Its spectrum fits perfectly the glow expected from a uniform blackbody at a temperature of 2.726 K.
  2. Its temperature and intensity are uniform to 1 part in 100,000 all over the sky.
  3. Tiny fluctuations in temperature are indicative of a slight nonuniformity of densities and temperatures in the early universe.

C. This radiation comes from the moment when the first atoms formed, 380,000 years after the big bang.
  1. Prior to this time, the entire universe was too hot (>3000 K) for hydrogen atoms to form, since collisions between atoms and photons would immediately knock away any electron that tried to attach to a proton and make ^{1}H.
  2. With all the low-mass, speedy electrons running around freely, photons could only travel a short distance before colliding with one and changing direction.
  3. When further expansion cooled the universe slightly below 3000 K, hydrogen atoms could form and, with all the electrons quickly binding to atoms, the photons were allowed to flow freely in a straight line—directly, 13.7 billion years later, to our telescopes, where we observe them as the CMB.

D. Modeling this radiation provides us with the most accurate estimates of the parameters that describe the big bang model.

Suggested Reading:

Harrison, *Cosmology*.

Rees, *Our Cosmic Habitat*.

Questions to Consider:

1. Is the CMB today the same here and 12 billion light-years from Earth?
2. How much federal funding should go toward making ever-more exquisite measurements of the properties of the CMB?

Lecture Twenty-Three—Transcript
The Limits of Vision—13.7 Billion Years Ago

Relativity has provided 2 critical results that we'll need in our search for the ultimate origin of matter and the beginning of history. First, that mass and energy are equivalent and interconvertible. A photon with 0 mass, a packet of electromagnetic energy, can spontaneously split into 2 particles, a particle and an antiparticle, both of which have mass. And vice versa: The combination of particle and antiparticle can eliminate that mass from the universe and turn it into pure energy. Secondly, that space and time are inextricably linked. Looking out into space is, quite directly, looking back into time. To read the history of the universe, we need only turn the pages by looking deeper and deeper into space.

We are ready, then, for the final quest: to find the very first atoms, to learn of their history, and then to push on to the creation of the electrons and the quarks from which they are made. As will become clear, we cannot see the creation event directly. But clues, written in the photons which we can see, allow us to tell a remarkably precise story about what must have gone on in the first microsecond of the universe.

It was 80 years ago, in 1929, when we first recognized that the Milky Way was not the universe. This achievement was the result of the work of 2 men, one of whom you have likely heard of and the other of whom you almost certainly have not: Edwin Hubble, who is familiar because of his eponymous telescope, and Vesto Slipher, who languishes in undeserved obscurity. Slipher was another of the young astronomers, like Douglass of the tree-ring fame, hired by Percival Lowell to nightly chart the progress of the advanced civilization he thought existed on Mars, digging away at their canals. Like Douglass, Slipher had very little interest in Mars and set out on another project of his own: to measure the motions of the spiral nebulae.

In addition to the billions of stars one sees in a telescope scattered across the sky and the dark clouds where new generations of stars are formed, as I've shown you, there are little spiral pinwheels of light that exist throughout the sky. In the early part of the 20th century, these were, in fact, thought to be solar systems in formation, a not unreasonable idea in that I told you real solar systems do form from collapsing spinning clouds which end up with a ball in the middle, the star, and a thin disc around them which make the planets. These little pinwheels looked like they might be solar systems in formation. But it wasn't clear, and Slipher set out to see if he could measure their motion and if that would tell us something about their origin.

Waiting for them across to move across the sky in a lifetime would never have yielded anything of interest. Instead, Slipher looked at the motion towards or away from the Earth. He did this by using an effect which is familiar to you, although the name may not be, the Doppler effect. The fact that any object emitting a wave, when it comes towards you has the waves squished together—and, therefore, for sound, raises the pitch; for light, shifts the light to the blue—and then as it moves away from you again, spreads out the wave peaks such that for sound, the sound is lower, and for light, the light is redder.

You've all been familiar with this from hearing a truck go by on a highway or an ambulance approaching and then receding. The sound is characteristically high pitched when it's coming towards you, because the waves are squished together and the distance between the peaks is small, and lower as the waves are spread apart as the object disappears. I can demonstrate that with this little beeper here. As I spin it around, you'll hear the pitch change, high when it's coming towards me and low when it's going away. That's called the Doppler effect, and it works for light just as it does for sound. The light squished closer together means wavelengths shifted to the blue; the light waves stretched farther apart means wavelengths shifted to the red.

To measure this, one can't just simply take a picture and see if the spiral nebula is blue or red. The shifts, for reasonable velocities, are tiny. What one has to do is take advantage of the unique bar code of spectral lines of, for example, hydrogen, with photons whose energy is known precisely because they correspond to the jump of an electron from one orbit to another. This means it's painstaking work. In the early part of the 20th century, photography was well developed, but photographic film was a really poor recorder of light. The reason we've all shifted to digital cameras these days is because photographic film typically converts about 1% of the light falling on it into a signal, whereas digital cameras operate at an efficiency closer to 90%.

These spiral nebulae were very faint and so Slipher had to carry his photographic glass plate out to the telescope—in the dark, of course—slide it into the camera; expose all night, carefully moving the telescope as the object tracked across the sky; and then carefully take the plate out in the morning; make sure it was covered in black paper; take it back to the darkroom and store it; and come back the next night to repeat the measurement. Typically, it would take a week of a single exposure to get a measurement of the position of the line to measure the Doppler effect. In fact, between dropped plates, and cloudy conditions, and the work that he had to do for Lowell looking at Mars, it took Slipher nearly 15 years to get 21 good measurements. But the result was quite remarkable: 17 out of the 21 spiral nebulae he observed were moving away from us. That is, their lines were shifted to the red. This is really quite astonishing if you imagine a universe in which things are randomly moving around. It's like getting 17 heads out of 21 flips of coin; the odds against that are nearly 100,000 to 1.

Enter Edwin Hubble. He was focusing on trying to measure the distances to these same spiral nebulae. He had, in fact, using the new large telescopes on Mount Wilson outside of Los Angeles, resolved individual stars in these systems and no longer believed that they were clouds of gas but separate galaxies, as he called them, other island universes like our own, expanding enormously the scope of the universe. The distance he calculated to each galaxy was based on measuring the pulsations of stars whose intrinsic luminosity he knew from studying them in the Milky Way. These stars expand and contract slowly with time, as I've mentioned, with periods ranging from weeks to months. It turns out that the amount of time they take to go from maximum to minimum and back again is directly proportional to their intrinsic luminosity. And so having determined this in our galaxy from nearby examples, he then used the same kinds of stars in other galaxies to measure their distance. Combining his results with Slipher's, he discovered a remarkable result: The farther away a galaxy was, the greater its red shift, the faster it was moving away.

Here's Hubble's data. We plot on the y-axis the velocity of the galaxy in kilometers per second. Negative velocities, by convention, mean objects moving towards us; positive velocities are velocities moving away. We see that there are a few galaxies, a handful, 3 or 4, that are moving towards us at a few hundred kilometers a second. Indeed, the nearest large galaxy to our own, the Andromeda galaxy, and our galaxy are moving towards each other at about 350 kilometers a second, and a few billion years from now, they'll actually collide. Colliding galaxies aren't quite as drastic as they sound, because the star separation, remember, is enormous. The orange here in Washington and the orange in Minneapolis—if you throw a bunch of oranges separated by that distance from each other, they're very unlikely to hit head on. The galaxies will basically pass right through each other, but the orbits of the stars will indeed be messed up.

However, if we look at the majority of the galaxies in Hubble's data, we see they're moving away, from hundreds up to more than 1000 kilometers a second. Furthermore, by plotting their distance from Hubble's measurements on the x-axis, we see that there's a correlation between these 2 quantities; that the nearby galaxies are the only ones that are coming towards us or moving away at modest speeds, whereas the farther we go away, millions of light-years out into space, the faster the galaxies are running away—even more astonishing.

The naïve interpretation of Hubble's observation is very disturbing. If all the galaxies are running away from us, it would seem that we must be the center of the universe. Furthermore, if the farther away they are, the faster they're fleeing, then they must know we're here and where we are. Both these thoughts violate something we call in cosmology the cosmological principle. That principle is that we do not occupy a special place in the universe. Now, as a human-generated notion, the cosmological principle could, of course, be incorrect. Nonetheless, it's worth trying to think of alternative explanations for Hubble and Slipher's extraordinary discovery of the galaxies fleeing from us.

The simplest alternative is that spacetime itself is not static but that the space between the galaxies is expanding. If spacetime is stretching, the distances between all pairs of galaxies will increase. We don't need to be the center; any galaxy you sit on will see all the galaxies running away from it. Furthermore, the greater the original distance between 2 objects, the faster the distance grows as the spacetime expands uniformly in between.

The galaxies arrayed on the surface of a balloon illustrate this effect. Here I have a little nighttime black balloon with a bunch of little galaxies affixed to it. As I blow up the balloon, or expand the spacetime, you see that the galaxies, which are barely separated by a finger width now, move apart. Now, they're separated by up to 3 finger widths. And the 2 galaxies that were initially furthest apart moved apart faster still. Now, they're back together when the universe shrinks.

Let's see how this works in 1 dimension rather than doing a 2-dimensional model like this, because it's confusing enough as it is. The observation of Hubble and Slipher is that we sit in a place where all the galaxies are running away from us. Let's take a 1-dimensional analogy here. Imagine a ruler made of rubber that's 12 inches long and affixed to it are 4 galaxies separated by different distances from each other. Now let's imagine we take the end of that ruler and we stretch it. We stretch it at the rate such that it doubles every 30 seconds, or expands by 24 inches per minute. Let's look at what happens to each pair of galaxies.

The galaxy at the origin of the ruler, of course, stays at the origin of the ruler. That will be, say, our galaxy. And so its point has no velocity and is plotted at a distance of 0. The next galaxy out, which is located at the 2-inch mark on the ruler, begins life 2 inches away, but as we stretch the ruler, ends up, after 30 seconds, 4 inches away and, after a minute, 8 inches away. It's moving away at 4 inches per minute when it initially started 2 inches from us. That gives us a second point on this graph of velocity versus distance apart.

Now let's look at the third galaxy. It starts out life 5 inches away from us, and as we stretch the ruler, that distance goes to 10 inches, to 15 inches, to 20 inches, as we keep doubling the size of the ruler. It's moving apart at 10 inches per minute when it started out 5 inches away from us, a third point on the graph. Finally, the last galaxy, which starts out at the opposite end of the ruler, begins life, of course, 12 inches away and then, as we double the size of the ruler, goes to 24 inches away. It's moving away at 24 inches per minute, having started at 12 inches from us, a fourth point on the graph.

Now, of course, we connect the dots. The line on the graph rises from the lower left to the upper right, just like Hubble's galaxies did, which we interpret as saying that the velocity with which they're moving away increases with their distance from us. This looks like a nice straight line, and you may recall from ninth-grade algebra that the equation for a straight line is that $y = mx + b$, where y is the value of the y coordinate; x, the value of the x coordinate; b, the intercept with the y-axis, which here is 0, so we can ignore; and m, the slope of the line, the slope being calculated by the change in the y divided by the change in x. Here, we have 12 inches per minute on the y-axis and 6 inches on the *x*-axis, giving us a slope of 2 inches per minute per inch. This is effectively Hubble's law.

In Hubble's case, we replace the *y*-axis with the velocity of the galaxy in kilometers per second. The distance on the *x*-axis is the distance in millions of light-years, and the slope, which we name H in honor of Hubble himself, gives us the rate at which galaxies are moving away: $v = HD$, where H is roughly 20 kilometers per second per million light-years. That is, for every million light-years I go out into the universe, a galaxy is moving 20 kilometers per second more away from us. The space is uniformly stretching in all directions, or as we usually say, the universe is expanding. Thus, knowing that $v = HD$, we measure H using nearby galaxies, as Hubble initially did. We measure the velocity by using the Doppler effect, which is a very straightforward observational technique, and we can find the distance equals v/H, 2 quantities we know, the distance of any galaxy.

This result was extremely ironic for one Albert Einstein. His 1916 model of general relativity was dynamic; it had spacetimes that either expanded or contracted willy-nilly. And he was very concerned about this, so he added a fudge factor, called the cosmological constant, to make the universe static. Once it was discovered that the universe, indeed, was expanding, Einstein referred to this as his "greatest blunder." Numerous independent lines of evidence confirm that the universe is, indeed, expanding from a much denser phase in the very distant past. This is our framework for interpreting all of our observations and is called the big bang model.

The red shift induced by expansion provides a handy way to measure the distance of even extremely remote objects. Since all hydrogen atoms in the universe are identical, a theme I've been repeating throughout this course, the transitions of their electrons between energy levels lead to well-known emitted wavelengths. For example, the jump in the hydrogen atom between level 2 and level 1 corresponds to an energy difference of 10.4 electron volts, as we saw in Lecture Three, which produces a wavelength of light of 121.6 nanometers in the ultraviolet part of the spectrum—in the ultraviolet spectrum unless, of course, that light is red shifted.

By measuring the observed wavelength, which is observationally straightforward, we can tell how rapidly the galaxy is moving away from us and, therefore, using Hubble's law, how far away it is. The ratio of observed to emitted wavelengths gives the factor by which the universe has expanded since the light left that emitting galaxy. As the waves travel through space and the space itself expands, their crests grow farther and farther apart, as you can see on my white balloon here. We have a little wave which, if it's moving through a spacetime that's stationary, will always have the same distance between its crests. But as we stretch the spacetime, the distance between the peaks of the wave gets wider and wider. That, in effect, is the red shift. It's not, technically speaking, the Doppler effect, something moving through space; it's the space itself stretching, rendering the wavelengths longer and longer.

Having measured the rate of expansion using objects of known distance, the red shift thus directly gives the object's distance. And the most distant galaxies we have detected are at a red shift factor of 6.5. That means the radius of the universe, since the light left those objects, was about 7 times smaller. The universe was $7 \times 7 \times 7$, this dimension, this dimension, and this dimension, or 350 times denser at that point, because as we shall see, the number of fundamental particles has not changed since that time. The light left these galaxies 12.9 billion years ago.

How can we use this ancient light to learn more about the history of atoms? A few minutes after the big bang, only 6 types of nuclei were present in the universe, 2 isotopes each of hydrogen, helium, and lithium: normal hydrogen, a single proton, and heavy hydrogen, deuterium, a proton with 1 neutron; light helium, helium-3 that we encountered first in the nuclear processes that power the Sun, 2 protons and 1 neutron, and normal helium, alpha particle, helium-4, 2 protons and 2 neutrons; and finally, lithium, the third element in the periodic table, which has either 3 protons and 3 neutrons or 3 protons and 4 neutrons. Nothing else was present because nothing with more neutrons and protons had a chance to form.

The relative amounts of each of these isotopes provide an extremely strong constraint on the earliest moments of the universe. Measuring their abundances from a distance of nearly 13 billion light-years is certainly a challenging task. At this time, most of the matter had not yet collapsed into galaxies and was spread more or less uniformly through space in tenuous clouds. The initially densest regions, however, had strong enough gravity to be able to have collapsed, and some produced giant black holes, a billion times the mass of the Sun, which gobbled up all the surrounding material at a prodigious rate.

This is the first mention of black holes in 23 lectures. When I give public talks, the first question, no matter what I speak about, is usually about black holes. But you'll have save those questions for another lecture series, because we don't have time to talk about them now. Here, all we need to know is that they're massive objects which gobble up matter and in the process shine brightly. Yes, it sounds ironic: We call them "black holes" and light can't escape from them, but as the matter spirals into them, it heats up to very high temperatures and before it plunges over the edge of the horizon of the black hole, it emits enormous amounts of light. In fact, one of these black holes can easily generate 1000 times more energy than a galaxy of a 100 billion stars and so can easily be seen across the vast distances involved.

Most galaxies today have these black holes, so we're not surprised to find them as relatively common in the early universe. Our galaxy has a particularly wimpy one—it's only about 3.6 million times the mass of the Sun—but most galaxies have masses of black holes between 100 million and a billion times the mass of the Sun, and indeed, these holes play a major role in the development and evolution of galaxies.

In any event, as the shining light from these black hole beacons passes through the clouds of gas along the way, deuterium, heavy hydrogen atoms, absorb that light at the precise wavelengths corresponding to their electron jumps, which differ ever so slightly from those of normal hydrogen because of the difference in the mass of the nucleus. Again, not because gravity is involved, but because as the electron orbits the nucleus and the nucleus orbits in turn, the heavier one moves less than the light one and so the energy levels are shifted ever so slightly, producing different wavelengths of light.

Detecting both normal hydrogen and heavy hydrogen in the same cloud allows us to calculate their abundance ratio. Since this gas is unpolluted by the effects of nucleosynthesis in stars—indeed very few, if any, stars have formed yet—it represents the value of the primordial material formed in the big bang itself. And the answer is 30 deuterium atoms for every million hydrogen atoms. Deuterium is relatively rare, for a reason we'll explain next time.

The other 4 isotopes are much less abundant than hydrogen, deuterium, and helium, and they're very difficult to measure at high red shift, so their determinations are made more locally. Galaxies which have yet to experience much star formation, which have taken very little of their gas and processed it through stars, which create heavy elements, are relatively pristine samples of primordial material. Also, the very oldest stars in our own galaxy, the first generation to form—some of which are still around because the small stars, as you recall, live a very long time—are likewise made with material very close to that original composition.

From these observations, the primordial abundances appear to be as follows: Helium-4, what I've said from the start is the second most abundant element in the universe, at about 22%; helium-3 at about 50 parts per million; lithium-7 at only 3 parts per billion; and lithium-6, still undetected, at less than 1 part per trillion. Our instruments are just not sufficiently sensitive yet to see this tiny amount of expected lithium-6. These abundances are key to constraining the conditions in the first 3 minutes of the universe. Just as in the cores of stars, where the other elements are made, the number of each type of atom created is extremely sensitive to the temperature and the density in the nuclear cauldron where nuclei are created. We shall see next time how this works in the big bang.

The oldest photons in the universe are not from these gobbling black holes, back 12.9 billion years in time, but are even older than that and provide us with a direct picture of the structure of the universe and the conditions which obtained when it was a mere 380,000 years old. The most distant individual light sources, these black holes, have been discovered to lie at about 12.9 billion years away. When that light was emitted, the universe was roughly 850 million years old. It may well have taken that long for the diffuse gas to collapse due to gravity and form objects, these massive black holes, and stars bright enough to see.

The initial stars, I might note, were quite different than stars in our galaxy today. Because of their lack of any heavy elements, of any ion other than hydrogen and helium, the pressure inside the star of the light flowing out—which today limits stars to a maximum mass of around 80 solar masses, lest the star be blown apart by the energy radiating from its core—the lack of things to push on, heavy elements, in the early stars meant they could grow to 200, 300, 500, maybe 1000, maybe 2000 solar masses. So they were impressive stars indeed. We've never seen one of these individually yet, although the James Webb Space Telescope, to be launched in the next decade, has one of its primary goals to find these primordial stars. But the black holes are there.

In any event, those individual objects come from a time when the universe was hundreds of millions of years old. However, observing the sky at microwave wavelengths, at wavelengths of electromagnetic radiation ranging from a millimeter to a centimeter, one can see a uniform diffuse glow coming from every part of the sky—not individual galaxies or individual stars, but a uniform background of glowing bright sky. It is never nighttime in the microwave part of the spectrum.

We call this glow the cosmic microwave background, and analyzing its spectrum gives us the remarkable result that it fits perfectly to the glow of a uniform blackbody, just a warm object, like a star or the oven in your house, at a temperature of precisely 2.726° above absolute zero. Its temperature and intensity are uniform to 1 part in 100,000 over the entire sky. To picture 1 part in 100,000, picture the Empire State Building, which is about 1000 feet tall, and an ant standing on top of the antennae of the Empire State Building. The difference between those 2, which you could hardly notice with your naked eye, is the maximum to the minimum peak of the microwave background.

Remarkably uniform, but there are tiny fluctuations in temperature which are indicative of a slight non-uniformity of the densities and temperatures even earlier in the universe. Here it is: the baby picture of our universe. Photons emerging from a time when the universe was only 380,000 years old—which may sound like an old baby to you, but when you're 13.7 billion years old, that's really from your youth. The color codes here represent its slight temperature changes: Blue is the coolest spots; red is the hottest spots. It looks very mottled—that direction may be red; that direction may be blue—but again, the fluctuations are miniscule. They've been magnified here with the color scale; they're only 1 part in 100,000.

This radiation came from the moment when the first atoms formed, a mere 380,000 years after the big bang. Prior to this time, the entire universe was too hot—more than 3000°, the temperature of the surface of a star—for hydrogen atoms to form. That is, for protons to capture electrons and keep them bound in orbits, since collisions between the atoms and the photons that were running around everywhere would immediately knock any electron away again that tried to attach itself to a proton and make hydrogen. With all these low-mass, speedy electrons running around freely, photons couldn't travel very far before they bumped into one and changed direction.

When further expansion cooled the universe as the stretching of spacetime shifted the peaks of the wavelengths apart so the temperature effectively fell just below 3000°, the hydrogen atoms could form and the photons were no longer energetic enough to knock them apart. All the electrons quickly bound up to atoms, and the photons were allowed to flow freely in a straight line through space directly to us, 13.7 billion years later, captured by our telescopes, where we observe them as the cosmic microwave background.

If you go outside at the end of this lecture and turn your face to the sky, 1000 trillion of those cosmic microwave background photons will hit your face every second. You will literally be bathing in the glow of the big bang. Analysis of this radiation provides us with the most accurate estimates of the parameters that describe the big bang model. They give us the age of the universe as 13.72 ± 0.12 billion years, a measurement of the age of our cosmos accurate to $\frac{9}{10}$%—almost as good as Bishop Ussher. It tells us the amount of matter in our universe is 4.56%, with an uncertainty of only $\frac{3}{10}$% of the total composition of the universe.

What? The matter I've been telling you about for 23 lectures makes up only 4.5% of the universe? I'm afraid that's true. It's a secret I've been holding from you: 95.5% of the universe is made of something else. What? We don't know. We're ignorant. We know it's of 2 types, and to cover our ignorance, we call one dark matter and the other dark energy. Dark matter makes up 22.8%, dark energy, 72.6%; they are the subject for another course. And finally, the microwave background gives us the Hubble constant to supreme accuracy also. It's 21.6 ± 0.4 kilometers per second per megaparsec.

We have now seen directly, then, in the cosmic microwave background, when the first atoms formed. We are but 1 step removed from our final goal, to discover how and when the building blocks of atoms—electrons, protons, neutrons, and quarks—came into existence. Using the information from the primordial element abundances and the cosmic microwave background, we are now prepared to go beyond where we can see directly to the first moment of the universe.

Lecture Twenty-Four
The First Few Minutes—Where It All Began

Scope:

Prior to recombination, the universe was too hot for electrons to stay attached to their nuclei. The speedy electrons of that era collided frequently with light, scattering it in all directions. But from the moment space was cool enough for atoms to form, the light streamed directly to us, carrying with it messages from earlier times. Examination of tiny fluctuations in the temperature and intensity of that light across the sky allow us to infer the conditions of the first microseconds of the universe, when the quarks came together to make protons and neutrons and, a minute later, when the neutrons and protons stuck to make atomic nuclei. Following a single quark from 1 microsecond to today renders our atomic tour of history complete.

Outline

I. Although a direct view of the early universe is blocked by the hordes of free electrons that existed prior to the formation of the first atoms 380,000 years after the big bang, a host of constraining observations nonetheless allow a precise model of the universe's birth to emerge.

II. Our journey to the beginning of time will conform to all the rules of physics we have learned.
   A. Mass and energy are interconvertible.
      1. Photons can make particle-antiparticle pairs if they have enough energy to make up the sum of the masses of the 2 entities.
      2. Particle-antiparticle pairs come together and annihilate their mass into photons.
   B. A photon's energy is inversely proportional to its wavelength.
      1. Higher energy means shorter wavelengths.
      2. As the universe expands, wavelengths of photons are stretched, thus lowering their energies.
   C. These 2 concepts lead to "freeze out": While particle-antiparticle pairs can always make photons, photons stop making pairs when their wavelengths get too long.

III. Running the clock backward from the 380,000-year mark from whence the CMB emerged, the universe continues to get smaller, hotter, and denser.
  A. At 380,000 years, the conditions are very well established by the characteristics of the CMB.
   1. The temperature is 3000 K.
   2. There are about 1250 protons and electrons in every cubic centimeter of space—roughly the density of today's star-forming clouds in the interstellar medium.
   3. There are roughly 1 billion photons for every proton as a consequence of events at the very earliest times.
   4. The number of particles and the number of photons have been roughly constant since then; the only difference is that the photons have cooled and the particles have spread out with the billionfold expansion of the universe.
  B. Indeed, running the clock all the way back to when the universe is just 3 minutes old, nothing changes except that the temperature rises and the density increases.
  C. Between the age of 1 second and 3 minutes, however, there is lots of action.
   1. At $t = 1$ second, the temperature is 10 billion K and the density of matter is $\frac{1}{10}$ that of water everywhere.
   2. This era begins with 2 neutrons for every 10 protons because of earlier conditions we will explore shortly.
   3. Neutrons and protons readily collide and stick to make deuterium.
   4. However, the photons at this temperature are energetic enough to split deuterium apart again.
   5. Simultaneously, neutrons—which, when they are free, decay to a proton and an electron with a half-life of 10.3 minutes—start disappearing.
   6. After 90 seconds or so, there are 14 protons for every 2 neutrons, but the temperature has fallen enough that the photons can no longer split deuterium and all the neutrons combine with protons, leaving 12 protons and 2 deuterium nuclei for every 16 particles.
   7. Deuterium nuclei are highly reactive at these temperatures, and they quickly bind together to form helium-4.

8. After 3 minutes, it is all over; we are left with 1 helium-4 nucleus for every 12 protons (i.e., the observed 25% of the mass in helium), plus trace amounts of deuterium, helium-3, and lithium.
9. The exact amount of helium-4 produced is very sensitive both to the temperature during this time and to the initial proton-to-neutron ratio.
10. The amount of deuterium left over is very sensitive to the density of matter at this epoch.
11. Thus the observational constraints on these abundances we described last time give us considerable confidence that we understand what was happening.

D. To discover why the initial proton-to-neutron ratio was 10 to 2, we need to go back to the lepton era, between $\frac{1}{10,000}$ of a second and 1 second after the big bang.
 1. This era began with the temperature at 1 trillion K and a matter density of 0.1 tons/cm^3, or the density of me squeezed into a teaspoon.
 2. Because of yet-earlier conditions, there were 1 billion photons, electrons, positrons, neutrinos, and antineutrinos for every proton.
 3. Neutrons combined with positrons to make protons, and protons combined with electrons to make neutrons.
 4. Since the neutron is 0.5% heavier than the proton, it takes slightly more energy to create it; near the end of the era, this small energy difference became crucial, and we end up with 5 times as many protons as neutrons (10-to-2 ratio).

E. To discover where these protons and neutrons came from, however, we have to go a step further back, to the hadron era.
 1. At 2 trillion K, photons have enough energy to spontaneously produce a proton and an antiproton.
 2. Since photons produce a particle for every antiparticle and particles and their antiparticles combine to produce photons, it is not clear why there is any matter left in the universe at all, since all the matter and antimatter should have combined and annihilated.

3. One microsecond (10^6 s) after the big bang, the temperature was 20 trillion K and proton/antiproton (and neutron/antineutron) pairs were created in great abundance.

F. Prior to 1 microsecond, the density was more than a billion tons per teaspoon, which is the density of nuclear matter.
   1. This means all the protons and neutrons in the universe were touching and overlapping each other, dissolving them into their constituent quarks.
   2. Somehow, at the time of this critical transition point where protons and neutrons solidified from the quark soup, a slight asymmetry left us with a billion and 1 protons for every billion antiprotons and a billion and 1 neutrons for each billion antineutrons.
   3. The matter/antimatter pairs annihilated, producing a billion photons, and the single proton and neutron went forward to form all of the matter in the universe.
   4. Perhaps the Large Hadron Collider will give us a hint as to why this asymmetry occurred; whatever it was, we wouldn't be here without it.

G. The quark era before 1 microsecond remains the subject of much speculation, but the story from that moment to today is pretty clear—and our history of atoms is complete.

IV. Since we are used to time running forward, let me review our microcosmical history by following a single up quark from its birth just before the 1 microsecond barrier.

A. At t = 1 microsecond, our up quark joins an up and down quark pair to form an unbreakable bond and become the universe's first proton.

B. At $t = 10^4$ s, it bumps into a frisky electron and joins with it to become a neutron, canceling out its positive charge.

C. At t = 10 s, bored in this neutral state, it spits out the electron to become a proton again.
   1. Within seconds, it bumps into a neutron and becomes a deuterium nucleus.
   2. A few seconds later, however, a photon comes along, splitting up the deuterium and knocking our proton back to its lonely state.

3. A minute later it finds another neutron companion, and this time the relationship lasts, since no photons remain powerful enough to tear it asunder.
4. Within seconds, it encounters another deuterium and quickly forms a nuclear family of 4—a helium nucleus.

D. Hundreds of thousands of years go by, but the helium nucleus remains positive.
1. By then, life has slowed down sufficiently that a couple of electrons wander by and are captured into orbits around the nucleus; the universe's first atom is complete.
2. Occasionally a passing photon will eject one of the electrons, but since all electrons are identical, a new one quickly replaces it and no one is the wiser.

E. Content with its 2 electrons, unwilling (indeed, unable) to form bonds with other atoms, our helium friend drifts around in an ever-expanding universe.
1. After a billion years or 2, however, it finds itself in an unusually dense part of space and is gravitationally drawn toward trillions of its brethren to help construct a galaxy, our Milky Way.
2. Visits to other atoms now require a trip of only a centimeter or so. It is part of an interstellar gas cloud.

F. After 7.5 billion years pass, our helium atom feels the neighborhood is getting crowded, with more than a million fellow atoms stuffed into its formerly isolated 1 cm^3 of space.
1. Things continue to get worse, and soon there are trillions of neighbors.
2. Its electrons get stripped for good, and it is back to its primal, positively charged state.
3. After 10 million years of constant jostling and ever-harsher conditions in the core of a massive star, it finds itself confronted simultaneously with 2 identical companions. In an instant, they snap together and form a lifelong bond—a carbon nucleus is born.
4. A few hundred thousand years later, our carbon finds itself thrust out of its steamy environment at nearly $\frac{1}{10}$ the speed of light by the explosive destruction of the star that made it, and soon Mr. C is in much less crowded circumstances.

G. A million years on, in frigid surroundings, our carbon bumps into a solid surface and sticks to it.
  1. On this cold little interstellar dust grain, there's a party; 5 other carbons, 2 nitrogens, 2 oxygens, and 14 hydrogens are invited.
  2. They are so attracted to each other, they make some bonds, in a left-handed sort of way at least, and form a lysine molecule.

H. Twelve million years later, the group lands with a thud on a kilometer-wide asteroid, where they chill out for a while.

I. Then, after several million more years pass, their temporary home falls hard onto the crust of the newly formed Earth.
  1. Things gradually settle down and cool off for our quark in the proton in the carbon nucleus in the lysine molecule, until one day, a few hundred million years later, our lysine finds itself floating in a pool of water.
  2. Eventually, a triplet of adenine molecules wanders by and, spying a lysine, grabs on tight.
  3. Our lysine soon finds itself unceremoniously glued to a whole bunch of neighboring cousins, amino acids all, making up a simple protein.

J. Now the fun begins—from the comfort of an archeobacterial cell wall to the sludge at the bottom of a pond; from the gut of a worm to the yolk of a bird's egg comes 4 billion years of constant transformation.
  1. Often our carbon atom breaks free from its companions.
  2. It spent millions of years alone, locked in the calcium carbonate of a Foraminifera shell 3 miles under the sea.
  3. Once it found itself shot out of the crater of a volcano, linked with 2 oxygens but otherwise free to roam the Earth.
  4. For 400 millennia, it was trapped in a tiny air bubble in Greenland's glacier, until at last it was liberated into a warm lab by a curious man with a needle.
  5. It escaped out the window but was soon captured by a tree leaf and linked again with dozens of companions into a molecule of cellulose.

6. Rousted out of a comfortable ring of wood by a man with a little borer and dropped unnoticed on the ground in a speck of sawdust, it was accidentally ingested by a passing pheasant pecking at the ground for seeds.
7. But its home in the pheasant's breast would also be short-lived, because I roasted that pheasant and ate it last week.
8. And now, with my final breath of this series, that original quark in our proton in our carbon nucleus, now joined with 2 oxygen friends, finds freedom again.

Suggested Reading:

Greene, *The Elegant Universe*.

Kirshner, *The Extravagant Universe*.

Questions to Consider:

1. Is it plausible that the laws of physics developed on Earth will apply without modification to the first microsecond of the universe?
2. What journey has that quark in the proton in the atom on the end of your nose been through?

Lecture Twenty-Four—Transcript
The First Few Minutes—Where It All Began

Last time, we found that our direct view of the early universe is blocked by the hordes of free electrons which existed prior to the formation of the first atoms 380,000 years after the big bang. With only $\frac{1}{2000}$ the mass of a proton they consort with, they are now moving 45 times faster—that's the square root of 2000—than the protons. And they intercept photons of light at every turn, scattering them hither and yon, obscuring a clear view to the beginning of time. This is because the universe is just too hot, and thus the photons too energetic to allow the protons to capture the electrons into stately orbits and get them out of the way.

So we cannot see directly farther back into the past than 13.7196 out of the 13.7200 billion years since the universe was born. That's still pretty good, but it's not all the way and won't allow us to fulfill our promise to measure things back to the beginning of the constituents of atoms. As the 17th-century French mathematician and philosopher Blaise Pascal said, in another context, "If our vision be arrested here, let our imaginations pass beyond." In our case, our imagination is highly constrained by the host of observations that allow us to construct a precise model of the universe's birth.

Running the clock backwards from the 380,000-year mark, the last thing we can see, the universe continues to get smaller, and hotter, and denser. Smaller, because we're going backwards in time; the spacetime is shrinking. Hotter, because as that spacetime shrinks, the photons that are waving through it have their peaks compressed, and shorter wavelengths mean hotter temperatures. And denser, because the number of particles in the universe remains constant for a long time, and as you squeeze the space for the same number of particles, it just is denser.

As we continue our journey to the beginning, we will pass through, first, the radiation era, where photons rule, and then, successively through the lepton era, the hadron era, and the quark era, where these particles, our trusted historians, had their birth.

There are a few physical processes which it is important to review before we embark on this journey to the beginning of time. (1) Mass and energy are interconvertible. Photons with sufficient energy can spontaneously create particle-antiparticle pairs—never only a particle or an antiparticle but particle-antiparticle pairs. And whenever a particle meets an antiparticle, the 2 come together and annihilate, usually into a massless photon. The equivalence is given by Einstein's famous equation with just a little twist: $E = 2mc^2$, where c^2 is the velocity of light and m is the mass of one of the particles. The 2 is because we always create a particle-antiparticle pair.

The second important physical principle is that a photon's energy is inversely proportional to its wavelength. That is, short wavelengths mean high energy; long wavelengths mean low energy. As spacetime expands, the photons traveling through it are stretched to longer wavelengths and, thus, lose energy.

Putting these 2 concepts together, we can introduce the idea of "freeze out." When photons are energetic enough, they freely exchange places with particles. A photon goes to, for example, a proton plus an antiproton, and a proton and an antiproton come together to make a photon. This goes on time after time after time.

In the ever-expanding universe, however, a moment will come when the photon energies fall below the threshold they need for creating a particular particle-antiparticle pair. The particle-antiparticle pairs around at that time can still annihilate and make photons. But when they make that photon, the universe then stretches a little bit, the peaks of the waves of the photon get a little farther apart, they lose energy, and they can no longer reconvert to particles. Any particle stuck without an antiparticle mate will be left in that state forever; it's sort of an extreme form of musical chairs. And every photon in that state is doomed, likewise, to travel at the speed of light through space for eternity.

The freeze-out temperature is different for particles of different masses. Protons obviously require more energetic photons to produce them than do electrons, because their masses are nearly 2000 times greater. These different freeze-out points for the different kinds of particles we make are the demarcation lines between the eras of the early universe.

Starting at [380,000] years, the last thing we can see, the conditions are very well established by the characteristics of the cosmic microwave background. The temperature is 3000° Kelvin. Tiny fluctuations around that value are present and reflect small energy fluctuations from much, much earlier times. As the universe evolves, these tiny inhomogeneities in the background grow through gravitational interactions to become massive galaxies. But for now, at this point, the matter and photons are really very smoothly distributed throughout space.

There are about 1250 protons and electrons in every cubic centimeter of space—every sugar cube has over 1000 protons and electrons—roughly the density of star-forming clouds in interstellar space today, the places where the Sun and other stars form. The whole universe, however, is still to that density.

There are roughly 1 billion photons for every proton. This is a consequence of events that occurred at the very earliest times, which we'll explore shortly. The number of protons and electrons and the number of photons has been roughly constant from then until now, 380,000 years in. The only difference is that the photons have cooled because the space has stretched so much, and the particles have spread out with the billion-fold expansion of the universe. Indeed, running the clock all the way back to when the universe is just 3 minutes old—from 380,000 years to 3 minutes—virtually nothing changes except that the temperature rises, as in our backwards-running clock, the universe shrinks and the peaks of the wavelengths come together, and the density increases, as we pack the same number of particles into a smaller and smaller space.

Between an age of 3 minutes and 1 second, however, there is lots of action. At $t = 1$ second—1 second after the birth of space, time, matter, and energy in the big bang—the temperature of the universe is 10 billion degrees Kelvin. And the density of matter is $\frac{1}{10}$ of that of water everywhere: the entire universe, $\frac{1}{10}$ the density of water. That's the matter. The radiation compressed into this tiny space with their wavelengths all shrunk is a million times more energy than the matter. It's a $\frac{1}{10}$ of a ton per teaspoonful—that's me in a teaspoonful, like that. This is the radiation era. The dominant forms of mass and energy, or mass-energy—remember, mass and energy are equivalent, so keeping track of the total is sufficient—the dominant species of this time is the photon.

When this era begins, there are 2 neutrons for every 10 protons, because of earlier conditions that we'll explore in a minute. The neutrons and the protons in this era readily collide with each other and they stick to make deuterium, heavy hydrogen, 1 proton plus one neutron. However, the photons at this temperature, when the universe is 1 second old, are energetic enough to split those deuterium atoms apart again. The binding energy of deuterium, the amount of energy given off when a proton and the neutron come together, as we saw in the cycle that powers the Sun, is about 2.2 million electron volts; that's how tightly they're bound. That's how much energy I must add if I want to see them split apart again. But at a temperature of 10 billion degrees, the energy of the individual photons is 4.3 million electron volts, so they have plenty of energy to knock apart any proton-neutron pair they come across. Thus, while deuterium forms, it's immediately split apart by the billion photons that exist for every one of the matter particles.

Simultaneously, neutrons decay. When a neutron is bound up in the nucleus of an atom, it can be stable for the life of the universe. But a neutron sitting alone on a table has a half-life—it's radioactive, if you will—that's 10.3 minutes. On average, a neutron will fall apart every 10.3 minutes. A collection of neutrons: Half will disappear in 10.3 minutes. And so, in the first couple minutes of the universe, the free neutrons running around, the ones that either failed to join with a proton to make deuterium or got knocked apart again by one of these overambitious photons, those neutrons are decaying, disappearing, spitting out an electron to turn the neutron into a proton. Thus, after 90 seconds or so, there are only 2 neutrons for every 14 protons, as opposed to 2 neutrons for every 10 protons when we started the era at 1 second. And with the expanding universe rapidly stretching the photon wavelengths, the temperature has fallen enough that the photons can no longer split deuterium. We have fewer neutrons around because some of them decayed. But all that exist quickly join with protons, because that reaction is very much favored, and photons can no longer split them. That leaves us with 12 protons and 2 deuterium nuclei for every 16 original particles. Again, the 2 key features are neutron decay and deuterium formation.

Deuterium nuclei are highly reactive, and at these temperatures, they quickly bind together, 2 deuteriums, to form helium-4. After 3 minutes, it's all over. We're left with roughly 1 helium-4 nucleus for every 12 protons. Four particles over a 16 total is about the observed 22% of the primordial mass that we said before is in helium. In addition, a few of the deuterons, a few of the deuterium nuclei, meet up with a single proton and make helium-3 or combine together in a more complicated way to make 1 of the 2 isotopes of lithium. Those 6 components, hydrogen, deuterium, light helium, heavy helium, lithium-6, and lithium-7, were the 6 primordial elements we discussed last time.

The exact amount of helium-4 produced is very sensitive both to the temperature during this time and to the initial proton-to-neutron ratio. Likewise, the amount of deuterium left over, the 30 nuclei out of every million that we found last time by measuring the light from distant black holes shining through the clouds of gas in between us and them, is very sensitive to the density of the matter at this epoch. So we have proton-to-photon ratio, we have temperature, and we have matter, all constrained by the measurements of these primordial isotopes. Thus, these observational constraints on abundances give us considerable confidence that we understand what was happening during this key time between 1 second and 3 minutes into the universe's life.

To discover why the initial proton-to-neutron ratio was 10 to 2, we need to go back, beyond the radiation era, to the lepton era, to the time between $\frac{1}{10,000}$ of a second and 1 second after the big bang. I know this is starting to sound a little absurd, that we can hope to understand what was going on at $\frac{1}{10,000}$ of a second after the beginning the universe, but follow the story; it hangs together.

In this era, we began with a temperature of about 1 trillion degrees, and the matter density, just the matter alone, was $\frac{1}{10}$ of a ton per cubic centimeter, me in a teaspoon again. Because of yet earlier conditions, there were 1 billion each of photons, electrons, positrons, neutrinos, and antineutrinos for every single proton, 1 billion to 1 for each of those fundamental subatomic particles.

Neutrons combined with positrons to make protons: a neutral particle, add a plus plus sign, you get a proton. And protons combined with electrons to make neutrons: a positive proton plus a negative electron combines, makes a neutral particle. This went on willy-nilly. Remember there was only 1 proton for every billion electrons, so there were plenty to join up with. There was only 1 neutron for every billion positrons, so there were plenty of positrons to join up with. This went back and forth, back and forth, over this entire period of $\frac{1}{10,000}$ of a second to 1 second. Since the neutron is $\frac{1}{2}$% heavier than the proton, just slightly heavier, it takes slightly more energy to create it. Near the end of this era, the freeze-out idea comes in, this small energy difference becomes crucial, and we end up with 5 times as many protons as neutrons, or this initial proton-neutron ratio of 10 to 2.

To discover where these protons and neutrons came from, however, we have to go a step further back, to the hadron era, in which the baryons are produced. Those of you with a good memory will recall from Lecture Two that baryons are triplets of quarks. The common triplets we know, protons and neutrons.

At 2 trillion degrees Kelvin, photons have enough energy to spontaneously produce a proton-antiproton pair. The energy of the photon at 2 trillion degrees is about 2 billion electron volts, and the mass of a proton $\times c^2$ is about 1 billion electron volts. So a 2-billion-electron-volt photon has enough energy to produce a 1-billion-electron-volt proton and a 1-billion-electron-volt antiproton. Photons can go to proton-antiproton pairs; proton-antiproton pairs go to photons.

One microsecond after the big bang, the temperature was 20 trillion Kelvin, and proton-antiproton and neutron-antineutron pairs were generated in great abundance. Since photons produce a particle for every antiparticle and an antiparticle for every particle, particles and their antiparticles combine to produce photons, it's not clear why there's any matter left in the universe today at all, since all the matter and the antimatter should have combined and annihilated, leaving only photons. Then, when the spacetime expanded enough so those photons could no longer produce particles and antiparticles, the universe should have been empty. It should have been photons alone.

Prior to 1 microsecond, the density was more than a billion tons per teaspoonful. That's the density, you may recall, of a neutron star, which is equivalent to the density of nuclear matter. The entire universe was a giant atomic nucleus. And this meant the protons and the neutrons were squeezed so close together they were effectively touching, as they do in a nucleus, overlapping with each other and dissolving into their constituent quarks.

Somehow, at this time, at this critical transition where protons and neutrons solidified from the quark soup and became distinct entities which were then carried forward to the universe today, a slight asymmetry left us with a billion and 1 protons for every billion antiprotons and a billion and 1 neutrons for every billion antineutrons. The matter-antimatter pairs annihilated, producing a billion photons, so that for every billion of the antimatter and matter pairs, only a single proton and neutron were left. These lonely remnants comprised all of the matter in the universe today. Perhaps the Large Hadron Collider, soon to come online in Switzerland, will give us a hint as to why this asymmetry occurred. Whatever it was, we wouldn't be here without it.

The quark era, before 1 microsecond, remains the subject of much speculation, but the story from that moment to today is pretty clear, and our history of atoms is now complete.

This was the goal we set at the outset, to construct all of history from the first microsecond of the universe to today. We've used all the actors of the micro world in this process: protons and neutrons to form the isotopes that mark chemical and geological processes; electrons jumping in their orbits to identify these isotopes and photons that signaled their presence across the universe and even to see the atoms form in the background radiation; positrons to signal radioactive decay; neutrinos to peer into the core of the Sun; and quarks to provide the yet-hidden asymmetry that allows matter to exist in the universe today.

This is the story as told backwards in time. But we're used, of course, to time running forwards. So let me review our microcosmical history by following a single up quark from its birth just before the first microsecond barrier.

At $t = 1$ microsecond, our up joins another up and a down quark to form an unbreakable bond and become the universe's first proton. At $t = 10^{-4}$ seconds, it bumps into a frisky electron and joins with it to become a neutron, canceling out its positive charge. At $t = 10$ seconds, bored in this neutral state, it spits out the electron to become a proton again. But within seconds, it bumps into a neutron and becomes a deuterium nucleus. A few seconds later, however, a photon comes along, splitting up that deuterium and knocking our proton back to its lonely state. A minute later, our proton finds another neutron companion, and this time, the relationship lasts, since no photons remain that are powerful enough to tear it asunder. Within seconds, it encounters another deuterium and quickly forms a nuclear family of 4, a helium nucleus. Hundreds of thousands of years go by, but the helium nucleus remains positive. By then, life has slowed down sufficiently that a couple of electrons wander by and are captured into orbits around the nucleus. The universe's first atom is complete.

Occasionally, a passing photon will eject one of the electrons, but since all electrons are identical, a new one quickly replaces it and no one is the wiser. Content now with its 2 electrons, unwilling—

indeed, unable—to form bonds with other atoms, our helium friend drifts around in an ever-expanding universe. After a billion years or 2, however, it finds itself in an unusually dense part of space and is gravitationally drawn towards trillions of its brethren or, to be more precise, 10^{68} other protons and helium nuclei, most of which are hydrogen cousins, to help construct a galaxy, our Milky Way. Visits to other atoms now require a trip of only a centimeter or so; helium is part of an interstellar gas cloud.

Seven point five billion years pass, and it feels the neighborhood suddenly getting crowded, with more than a million of its fellow atoms stuffed into a formerly isolated 1 cubic centimeter of space. Things continue to get worse; soon, there are trillions of neighbors. Its electrons get stripped for good, and it's back to its primal, positively charged state. After 10 million years of constant jostling and ever-harsher conditions in the core of a massive star, it finds itself confronted simultaneously with 2 identical companions, and in an instant, they snap together and form a lifelong bond. A carbon nucleus is born.

A few hundred thousand years later, carbon finds itself thrust out of its steamy environment at nearly $\frac{1}{10}$ the speed of light by the explosive destruction of the star that made it, and soon, our Mr. Carbon is in much less crowded circumstances.

A million years on, in frigid surroundings now, it bumps into a solid surface and it sticks. On this cold little interstellar dust grain, there's a party. Five other carbons, 2 nitrogens, 2 oxygens, and 14 hydrogens are invited. They are so attracted to each other, they make some bonds, in a left-handed sort of way at least, and form a lysine molecule, an amino acid. Twelve million years later, the group lands with a thud on a kilometer-wide asteroid, where they chill out for a while.

Then, after several million more years pass, their temporary home falls hard—bang—onto the crust of the newly formed Earth. Things gradually settle down, cool off a little bit for our quark in the proton in the carbon nucleus in the lysine molecule, until one day, a few hundred million years later, our lysine finds itself floating in a pool of water. Eventually, a triplet of adenine molecules wanders by and, spying a lysine, grabs on tight. It soon finds itself unceremoniously glued to a whole bunch of neighboring cousins, amino acids all, making up a simple protein. Now the fun begins.

From the comfort of an archeobacterial cell wall to the sludge at the bottom of a pond, from the gut of a worm to the yolk of a bird's egg, 4 billion years of constant transformation passes by. Often, our carbon atom breaks free from its companions. It spent millions of years alone, locked in the calcium carbonate of a Foraminifera shell 3 miles under the sea. Once, it found itself shot out of the crater of a volcano, linked with 2 oxygens but otherwise free to roam the Earth. For 400 millennia, it was trapped in a tiny air bubble in Greenland's glacier, liberated into a warm lab at last by a curious man with a needle. It escaped out the window from the lab but was soon captured by a tree leaf and linked again with dozens of its companions into a molecule of cellulose. Rousted out of a comfortable ring in the tree by a man with a little borer and dropped unceremoniously onto the ground in a speck of sawdust, it was accidentally ingested by a passing pheasant pecking at the ground for seeds. But its home in the pheasant's breast would also be short-lived, because I roasted that pheasant and ate it last week. And now, here it comes, that original quark in our proton in our carbon nucleus, now joined with 2 oxygen friends, finds freedom again. There it goes.

That quark has been on a journey for nearly 14 billion years. Science has been on its journey to discover this tale for a mere 400. Galileo knew nothing of atoms, but in 1609, his telescope gave us new worlds to think about. More importantly, Galileo gave us a new approach to the universe, the notion that it *is* a great book "written in the language of mathematics," as he said, and that we could understand this book if we learned that language. In short, that the universe, the *uni*ty of the di*verse*, is comprehensible.

In that spirit, we've applied physics to matters art historical and archaeological, chemical, geological, and astronomical in order to enrich and extend our view of history. A few topics have practical applications; the use of isotopes in medical diagnosis and the importance of past climate change to future climate prediction touch on 2 of the most important topics of our age. Others—knowing when Native Americans learned to plant corn or how the carbon in your fingernail was cooked up inside a star—will affect neither your longevity nor your brokerage account. Nonetheless, it is my hope that these tales of atomic detection work have enhanced your appreciation both of history and of science and have expanded your perspective on your place in the universe.

Timelines

Epochs
(Dated from the moment of origin of the universe)

before 10^{-6} seconds .... Quark era: Quarks are free.

10^{-4}–10^{-6} seconds .... Hadron era: Quarks bind into protons and neutrons.

10^{-4}–1 second .... Lepton era: Electrons and positrons annihilate each other.

1 second–380,000 years .... Radiation era: At the end of this period, atoms form.

after 380,000 years .... Matter era: Matter dominates the energy density of the universe.

(Dated in years before the present: bya, billion years ago; mya, million years ago; ya, years ago)

4.56–3.8 bya .... Hadean eon: Earth cools, its crust forms, and life emerges.

3.8–2.5 bya .... Archean eon: Oxygen-producing cyanobacteria emerge.

2.5–0.55 bya .... Proterozoic eon: Oxygen level in Earth's atmosphere rises; multicelled life evolves.

0.55 bya/550 mya .... Phanerozoic eon begins.

550–250 mya .... Paleozoic era: Land plants, trees, and animals emerge; dinosaurs evolve.

535 mya .... Cambrian explosion: Huge diversification in organisms.

250–65 mya .... Mesozoic era: Dinosaurs rule the Earth.

65 mya .... Cretaceous-Tertiary (K-T) boundary: Mass extinctions caused by an asteroid striking Earth.

65 mya–present .... Cenozoic era: Mammals dominate; continents take current forms.

125,000–11,000 years ago .... Last ice age: Sea levels drop 130 meters; ice sheets nearly 4000 kilometers from the North Pole.

6000–4000 years ago .... Copper Age: Humans develop metallurgy.

2100–1600 years ago .... Roman Empire.

1600–600 years ago .... European Middle Ages.

Events
(Dated from the moment of origin of the universe)

	Event	Technique (how we know)
10^{-6} seconds	Quarks fuse into protons and neutrons; Quark era ends and Hadron era begins.	Hubble's law, cosmic microwave background (CMB) fluctuations
10^{-4} seconds	Nearly all protons and neutrons annihilate with their antiparticles; Lepton era begins.	CMB photon ratio
1 seconds	Nearly all electrons and positrons annihilate each other.	Baryon/CMB photon ratio
3 minutes	^{2}H forms.	Quasar absorption lines
3 minutes	^{3}He, ^{4}He, and ^{7}Li form.	Abundances in oldest stars
380,000 years	Atoms form, CMB breaks free, and matter rules.	Details of CMB radiation
1 billion years	Milky Way galaxy forms.	Ages of oldest stars

(Dated in years before the present)

Time	Event	Technique
4.565×10^9 years ago (ya)	Nearby supernova triggers a gas-cloud collapse to form our solar system.	^{26}Al → ^{26}Mg in meteorites
4.56×10^9 ya	First meteorite material forms; our solar system is born.	^{87}Sr/^{87}Rb clock
4.55×10^9 ya	Meteorites and planetesimals condense.	^{128}I + ^{129}I → ^{128}Xe + ^{129}Xe
4.50×10^9 ya	Moon formed by the collision of a giant planetesimal with Earth.	^{147}Sm → ^{143}Nd decay
3.85×10^9 ya	Oldest crustal rocks form.	U and Th decay sequences.
3.8×10^9 ya	Life emerges on Earth.	^{13}C/^{12}C ratios in zircons
3.8–1.8×10^9 ya	Atmospheric O_2 begins rising and fluctuating.	Banded iron deposits of Fe+O_2
2.5×10^9 ya	First cells with nuclei emerge.	Steranes (derived from steroids) present in Australian shales.
600×10^6 ya	Last "snowball" Earth.	Zircon U → Pb dating of carbon and boron isotope-ratio anomalies and anomalous sedimentation patterns (e.g., "cap-carbonates").

525×10^6 ya	Cambrian explosion.	Fossil record
251×10^6 ya	Greatest mass extinction in Earth's history; 90% of all species disappear.	Fossil record
64.5×10^6 ya	Chixulub meteor wipes out the dinosaurs.	Iridium layer; ^{235}U and ^{238}U decay
3.5×10^6 ya	Isthmus of Panama closes off Pacific and Atlantic Oceans.	Sea floor cores
2.8×10^6 ya	Large-scale glaciers appear.	Sea floor cores
155,000 ya	Orbit variations trigger monsoon cycles.	^{40}Ar/^{39}Ar and ^{18}O/^{16}O ratios.
125,000 ya	Sudden sea-level rise as previous ice age ends.	Coral growth record
26,000 ya	Peak of Last Glacial Maximum.	Coral U-Th decay sequence
17,000 ya	Australian cave paintings created.	^{14}C dating of pollen in mudwasp nests
10,900 ya	End of last ice age.	Tree rings
5300 ya	Ötzi the Iceman lived and died in the European Alps.	^{14}C dating and isotope ratios
4900 ya	Today's oldest bristlecone pines sprouted.	Tree rings
4800 ya (2800 B.C.)	Stonehenge built.	^{14}C dating
4510 ya (2500 B.C.)	Egyptian pyramids built.	^{14}C dating

3881 ya (1872 B.C.)	King Sesostris of Egypt, witnesses a solar eclipse, giving us the oldest verifiable, calendrical date.	Solar eclipse cycle
3865 ya (1855 B.C.)	West Texas Pecos cave paintings created.	^{14}C dating of organic materials in paint
2700 ya (700 B.C.)	King Hezekiah builds a water tunnel in Jerusalem.	^{14}C and U-Th dating
1900 ya (A.D. 100)	Roman amphitheater in Mérida, Spain, built.	^{14}C dating of plaster
1600 ya (A.D. 400)	Orinoco Indians cultivate beans and maize.	^{13}C/^{12}C ratios
1430 ya (A.D. 580)	Elaborate Portuguese villas built in Roman style.	^{14}C in plaster
1400–600 ya (A.D. 600–1100)	Medieval Warm period; Vikings in Greenland.	Tree rings
900 ya (A.D. 1100)	Native Americans in Midwest learn to plant corn.	^{14}C dating and ^{13}C/^{12}C ratios
730 ya (A.D. 1280)	Cathedrals in Aland Islands, Finland, built.	^{14}C in plaster and tree rings
710 ya (A.D. 1299)	Anasazi civilization at Chaco Canyon collapses.	Tree rings and carbon isotope ratios
600–200 ya (A.D. 1400–1800)	Little Ice Age.	Tree rings
320 ya (A.D. 1690)	Tower in Newport, Rhode Island, built.	^{14}C in plaster
125 ya (A.D. 1885)	Spanish forger illustrates Medieval manuscripts.	Neutron activation

Half-Lives and Other "Clocks" Used in Dating

10^{-6} seconds	Muon decay, from our frame of reference.
10^{-4} seconds	Muon decay, from its own fast-moving frame of reference.
10.3 minutes	Free neutron → 1 proton, 1 electron, and 1 antineutrino.
2.6 hours	Manganese-56 → iron-56.
12.7 hours	Copper-64 → nickel-64 (61%) + zinc-64 (39%).
15 hours	Sodium-24 → magnesium-24.
26 hours	Arsenic-76 → selenium-76.
2.69 days	Gold-198 → mercury-198.
8 days	Iodine-131 → xenon-131.
27 days	Chromium-51 → vanadium-51.
47 days	Mercury-203 → thalium-203.
60 days	Antimony-124 → tellurium-124.
244 days	Zinc-65 → copper-65.
28.8 years	Strontium-90 → zirconium-90.
30.2 years	Cesium-137 → barium-137.
88 years	Plutonium-238 → uranium-234.
269 years	Argon-39 → potassium-39.
5730 years	Carbon-14 → carbon-12.
~12,000 ya	Current limit of dendrochronology.
~60,000 ya	Current limit of carbon-14 dating.
244,000 years	Uranium-234 → thorium-230.
730,000 years	Aluminum-26 → magnesium-26.
~800,000 ya	Current limit of ice-core dating.
15.7 million years	Iodine-129 → xenon-129.

~80 mya .... Current limit of
                                      ocean-sediment dating.
83 million years .... Plutonium-244 → thorium-236.
710 million years .... Uranium-235 → lead-207.
1.25 billion years .... Potassium-40 → argon-40 (89%) +
                                      calcium-40 (11%).
4.47 billion years .... Uranium-238 → thorium-234.
4.51 billion years .... Uranium-238 → lead-206.
14.1 billion years .... Thorium-232 → lead-208.
47 billion years .... Rubidium-87 → strontium-87.
105 billion years .... Samarium-147 → neodymium-143.

Glossary

action potential: The shift in the voltage of a neuron from its resting state of approximately −70 mV to a positive voltage in response to a stimulus. The stimulus, acting through specialized receptor molecules depending on its type, opens ion channels on the cell wall so that positive sodium ions rush into the cell. The action potential then propagates along the length of the neuron's axon to convey the signal to the next nerve cell.

adenine: The molecule $C_5H_5N_5$, which plays several important roles in the chemistry of life, notably as 1 of the 4 base pairs that define the genetic code in the DNA molecule.

alpha: One of the 3 radioactive "emanations" discovered at the turn of the last century, now known to consist of a helium nucleus with 2 protons and 2 neutrons.

amino acid: One of the 20 molecules that form the building blocks of all proteins in life on Earth.

annihilation: The merger of a particle with its antiparticle, which results in the conversion of the mass of the 2 entities to pure energy through Einstein's famous proportionality $E = mc^2$.

antimatter: Each type of subatomic particle has an antimatter twin that shares all but one of the particle's properties; e.g., an antielectron (or positron) has the same properties as the electron, except it has a charge of +1 instead of −1.

asteroids: Rocky bodies in orbit around the Sun composed of material left over from the formation of the solar system that did not accrete onto any of the planets. The majority of asteroids are located between the orbits of Mars and Jupiter, although they sometimes stray inward and collide with Earth, as most famously happened 64.5 million years ago, ending the age of the dinosaurs. Since they are remnants of the first material to condense in the protosolar nebula, their isotopic composition reveals clues to the origin and age of the solar system.

atom: From the Greek for "indivisible," the smallest unit of matter that retains that matter's essential character and establishes the basis for its interaction with other atoms. There are 92 naturally occurring types of atoms, each representing a unique element. Each atom is composed of 2 components: the positively charged nucleus, which contains more than 99% of the atom's mass, and a surrounding cloud of negatively charged electrons. The nucleus, in turn, is composed of 2 types of particles: protons, which carry the atoms' positive charge, and neutrons, of similar mass but chargeless. In a normal (neutral) atom, the number of protons and electrons exactly balances, giving the atom a net charge of 0. An atom's characteristics are determined by its number of protons, and all species, from 1 proton to 92 protons, are represented.

atomic mass: The sum of the number of nucleons (protons plus neutrons) in the nucleus of an atom (or of all the nucleons in a molecule). The atomic mass number is written to the upper left of the chemical symbol for the element. In all nuclear transformations, the atomic mass is conserved.

atomic number: The number of protons in an atom. The atomic number defines the chemical nature of an element; it is written to the lower left of the chemical symbol for the element.

banded iron formations: Geological formations consisting of alternating layers of iron oxide–rich sedimentary rocks and iron-poor rocks. These features are among the oldest rocks on Earth, with ages ranging back to greater than 3 billion years. They are thought to have been formed when iron dissolved in seawater combined with oxygen released by the blooming of cyanobacteria on the Earth's oceans, and they are the source of much iron ore today.

baryon: Any member of the class of subatomic particles consisting of 3 quarks bound together; protons and neutrons are the most common baryons in the universe today.

beta: One of the 3 radioactive "emanations" discovered at the turn of the last century, now known to consist of either an electron or its antimatter counterpart, the positron.

blackbody: A theoretical object that absorbs 100% of the energy incident on it (i.e., it is perfectly black). The distribution of wavelengths of electromagnetic energy emitted by such an object defines a blackbody spectrum, which has a peak intensity inversely proportional to the object's temperature. The radiation from stars is a close approximation to a blackbody spectrum.

Calorie: The energy necessary to raise 1 kilogram of water 1 degree centigrade. It is commonly used as a measure of the chemical potential energy stored in food; an adult human needs approximately 2000 Calories per day to maintain the body at its optimal temperature and to fuel its activities. (It is written with a capital C to represent 1000 calories, or 1 kilocalorie)

Cambrian explosion: The sudden blossoming of life forms that appears about 545–530 million years ago in the fossil record when the first large multicellular and shell-forming animals emerged, along with flowering plants. The explosion coincides with the breakup of a large single landmass (Gondwanaland) into separate continents, which increased the area of shallow seas at their margins. It remains a matter of debate as to whether all the new life forms appeared in a sudden burst or whether the preceding fossil record incompletely records their gradual emergence over a longer interval.

carbon dating: The use of the radioactive isotope of carbon, ^{14}C, to determine the age of an object. ^{14}C is produced in the atmosphere when cosmic rays strike nitrogen atoms in the air. Any chemical process that involves carbon incorporates ^{14}C atoms at a rate of approximately 1 atom per trillion compared to the stable isotopes ^{12}C and ^{13}C. As soon as the chemical process ceases (e.g., a living plant or animal dies, or plaster dries), carbon incorporation ends and the radioactive ^{14}C atoms begin to decay, with a half-life of 5730 years. By measuring the ratio of remaining ^{14}C to ^{12}C and ^{13}C atoms, one can estimate the age of the object with great accuracy back to roughly 50,000 years.

Chaco Canyon: A major center of the Anasazi civilization of the American Southwest, built and occupied between A.D. 850 and 1200. Located in northern New Mexico, at its peak it is estimated that the Chaco Canyon settlement housed 5000–10,000 people in an elaborate network of apartment complexes. Isotopic evidence suggests that food to sustain this population had to be carried from farmland up to 50 miles away.

CMB: *See* **cosmic microwave background**.

cochlea: The fluid-filled canal of the inner ear that transduces incoming sound vibrations to nerve impulses that are sent on to the brain.

cosmic microwave background (CMB): The afterglow of the big bang. The radiation we observe today has a perfect blackbody spectrum with a temperature of 2.736 K; it represents the photons produced in the big bang from the annihilation of particles and antiparticles. These photons were released to stream freely through the universe when electrons first combined with hydrogen and helium nuclei about 380,000 years after the big bang. Roughly 1000 trillion photons of this radiation strike your head each second when you stroll through the park.

cosmic rays: Very-high-energy particles (electrons, protons, nuclei) or photons generated by the Sun, in interstellar space, and even beyond our galaxy that constantly bombard the Earth. Collisions with air atoms produce showers of secondary particles (e.g., muons) and generate radioactive isotopes from stable atmospheric atoms.

cyanobacteria: Blue-green algae that were the first organisms to sustain themselves through photosynthesis. The oxygen they produced changed the composition of the Earth's atmosphere, oxidized all the iron in the Earth's crust, and led to the near extinction of the previously flourishing anaerobic organisms.

deuterium: The heavy isotope of hydrogen containing 1 proton plus 1 neutron.

Doppler effect: Named after a 19th-century Austrian physicist, this is the effect produced when the source of a wave and the observer of the wave are in relative motion. When the two are approaching each other, the wavelengths of the wave are compressed, leading to a higher pitch (in sound) or a bluer color (in light). When the two are receding, the distance between the wave crests is lengthened, leading to lower pitch or redder light.

dust: *See* **interstellar dust grain**.

electromagnetic spectrum: The rainbow of electromagnetic waves generated by accelerating charged particles, of which light represents 1 of more than 50 octaves. Radio waves; microwaves; infrared, visible, and ultraviolet light; X-rays; and gamma rays are the types of electromagnetic radiation (ordered from longest to shortest wavelength).

electromagnetism: One of the 4 fundamental forces of nature, involving the interaction of particles having the property of charge; like charges repel, and unlike charges attract. Electromagnetic forces govern the behavior of matter from the scale of the atom to the scale of mountains.

electron: The lightest of the family of fundamental particles called leptons, electrons are the negatively charged constituent of atoms; they have a mass of only 9×10^{-31} kg. Their specific arrangement around the atom's nucleus determines the modes in which the atom can interact with other atoms. When "shared" with another atom, an electron can form a bond that joins 2 (or more) atoms as a molecule. The linking and unlinking of shared electrons produces a chemical reaction in which one substance is transformed into another. Rearrangement of the locations of electrons in atoms and molecules are accompanied by the emission or absorption of electromagnetic energy.

electron volt: The unit of energy appropriate to describing interactions on the atomic and subatomic scales; 1 eV is the energy acquired by a single electron accelerated through an electric potential of 1 volt (1 eV = 1.6×10^{-19} joules). Chemical interactions typically release a few tenths of an electron volt to a few electron volts; electrons are bound to atoms with a few to a few hundred thousand electron volts; and nuclear interactions typically involve energies of a few to a few hundred million electron volts. The energy of a photon of visible light is about 1 eV.

element: One of the basic units of matter with definable chemical properties. Ninety-two different types of atoms (those containing from 1 to 92 protons in their nuclei) are said to occur naturally, although almost 2 dozen more have been created in the laboratory and are likely to be created in the explosions that mark the deaths of massive stars.

energy: A fundamental organizing principle of modern physics, energy represents the ability to do work—to make something move, stop moving, change direction, emit light, and so forth. Each of the 4 fundamental forces has an associated form of energy.

energy levels: The discrete amounts of energy, unique to each type of atom, with which an electron can be bound to its nucleus. Electrons can exist in each of the levels corresponding to its type of atom but in no others; transitions between levels can emit and absorb electromagnetic radiation.

escape velocity: The velocity an object must acquire to "not fall down"—to escape from its parent body. It is a function only of the mass and radius of the parent. The escape velocity for Earth is 11.2 km/s.

fission: The splitting, spontaneous or induced, of an atomic nucleus into 2 roughly equal pieces. For heavy nuclei (those containing more than 56 protons and neutrons), fission releases large amounts of energy. Nuclear power plants and submarines operate via controlled nuclear fission.

Foraminifera: Single-celled organisms that comprise a large part of oceanic plankton. They form calcium carbonate shells that, when they die, sink to the ocean floor, producing large limestone deposits. The isotopic composition of these shells, as well as the mix of species present, provide clues to the history of past climate.

freeze-out: The notion that once the energy of a photon in the expanding universe has its wavelength stretched (and thus its energy reduced) beyond a certain point, it can no longer produce particle-antiparticle pairs (because their combined mass times c^2 exceeds the photon's energy). At this moment, matter-antimatter annihilation can still occur to produce photons, but photons can no longer create matter, and the ratio of photons to particles is frozen. In the early universe, this happened in the first second.

fusion: The coalescence of protons and neutrons and/or nuclei to create heavier nuclei with, for nuclei with atomic masses less than 56, the concomitant release of energy. The stars are powered by nuclear fusion.

Galilean transformations: The equations that connect properties such as position, velocity, time, and mass in one reference frame with those in another reference frame that is moving relative to the first.

gamma: One of the 3 energetic particles that emerge from radioactive decay, now known to be a high-energy photon ($E > 100,000$ eV).

gas: A substance in which the constituent particles have relatively high individual kinetic energies and are thus not bound to one another, interacting like colliding billiard balls.

gravity: One of the 4 fundamental forces of nature, governing the interactions between particles that have the property of mass. Gravity is always attractive and dominates the behavior of matter on scales larger than several tens of kilometers (i.e., the Earth is held together by gravity but a mountain is not). The gravitational force is enormously weak, being about 10^{-39} the strength of the electromagnetic attraction between 2 oppositely charged fundamental particles.

greenhouse effect: The process through which a planetary atmosphere, while transparent to incoming radiation from the parent star at optical wavelengths (which heats the planet), is partially opaque to outgoing infrared radiation emitted by the planet, thus allowing the atmosphere to accumulate energy until its temperature rises to the point that the emitted infrared wavelengths become short enough to escape.

greenhouse gas: One of several gases that are efficient at absorbing outgoing infrared radiation from a planet, thus trapping energy in its atmosphere. The dominant greenhouse gases in the Earth's atmosphere are H_2O, CO_2, CH_4, N_xO, and chlorofluorocarbons.

hadron: Any particle consisting of quarks bound together by the strong nuclear force. In the universe today, there are 2 families of hadrons: baryons (consisting of quark triplets) and mesons (consisting of quark pairs).

hadron era: The era when the universe was between 10^{-6} and 10^{-4} seconds old and the hadrons emerged from the quark and lepton soup that characterized the earlier state of the universe.

half-life: The time in which the probability is 50% that a single radioactive atomic nucleus will decay, or equivalently the time in which half a sample of identical radioactive nuclei will decay. Half-lives range from fractions of a microsecond to tens of billions of years.

heat: The kinetic energy (energy of motion) of the atoms or molecules making up a substance.

H-R diagram: Named after the astronomers Henry Norris Russell and Ejnar Hertzsprung, who independently discovered its utility, the H-R diagram is a plot of the luminosity of stars versus their surface temperatures, which reveals patterns that led to our understanding of stellar structure and evolution.

hydrostatic equilibrium: The dynamic balance in a star between the inward pull of gravity and the outward push of thermal pressure sustained by nuclear reactions in the star's core.

ice core: A cylindrical piece of ice extracted from a glacier or ice sheet in which depth is directly correlated with time into the past. Ice cores from the Greenland and Antarctic ice sheets reveal the history of Earth's climate over hundreds of thousands of years through analysis of the isotopic composition of the ice itself, impurities deposited in the ice (e.g., sea salt and wind-blown dust), and the tiny bubbles of air trapped in the ice, which provide a record of atmospheric composition.

infrared: The portion of the electromagnetic spectrum just below ("infra") what our eyes perceive as red light. Infrared light is given off by objects ranging in temperature from 30 K to 3000 K.

interstellar dust grain: Particles ranging in size from large molecules to objects several microns in diameter and composed of silicates or carbon and iron compounds that permeate interstellar space. In dense regions of cold gas, dust grains provide the surfaces to which atoms from the gas phase adhere and form molecules.

interstellar medium: The material between the stars, composed of diffuse gas and small particles of dust.

ion: An atom or molecule that has either too few or too many electrons to be in charge balance. A negative ion has 1 or more extra electrons, while a positive ion is missing 1 or more electrons.

ion channels: Structures in the wall of a cell that open and close in response to various types of stimuli (heat, pressure, ion gradients, etc.) and allow ions to flow into and out of the cell. In nerve cells, they are responsible for initiating and propagating the action potential, which signals the brain when a stimulus has been recorded.

isotope: Atoms with identical numbers of protons and electrons that differ only in their number of neutrons. Since the number of electrons determines the atom's chemical behavior, all isotopes of an element behave identically in forming molecules with other atoms; the sole difference is their mass. This physical distinction can lead to their over- or underrepresentation in chemical processes. Hydrogen has 2 stable isotopes, one containing no neutrons and one containing a single neutron (deuterium), plus 1 unstable (radioactive) isotope with 2 neutrons (tritium). Radioactive isotopes provide a clock with which we can reconstruct history.

joule: A metric unit of energy, named after the 19th-century English physicist James Joule, corresponding to the energy of a 2-kilogram mass moving at 1 m/s. One watt is the expenditure of 1 joule of energy for 1 second; there are 4186 joules in 1 Calorie (kilocalorie).

Kelvins: The unit of temperature on the absolute (or Kelvin) scale, for which 0 is defined as the cessation of all motion. This zero point is equal to −273.15°C. The increments of the Kelvin scale are identical to those of the centigrade scale (with 100 units between the freezing and boiling points of water at sea level). While often referred to as "degrees Kelvin," the correct usage is simply "Kelvins."

kinetic energy: The energy contained in the motion of a body with mass. At speeds slower than the speed of light, kinetic energy is defined as $\frac{1}{2}mv^2$, where m is the mass of the object and v its velocity; as noted under joule, a 2-kilogram body moving at 1 m/s has an energy of 1 joule (1 J = 1 kg × m^2/s^2).

K-T boundary: The geologic and paleontologic feature that divides the Cretaceous and Tertiary geological eras (the K is from the German spelling of Cretaceous). The boundary marks the end of the reign of the dinosaurs and is now known to be a result of a massive asteroid impact with Earth 64.5 million years ago.

left-handed molecule: Some molecules come in 2 varieties that are chemically identical (they contain the same number of each kind of atom), but one version is the mirror image of the other. They are called left- and right-handed molecules because they rotate the plane of polarization of light passing through them to the left or to the right. All the amino acids that have a handedness in living things on

Earth are left handed, and there is a preponderance of left-handed amino acids in meteors as well, suggesting the solar system has a bias toward left-handed molecules.

length contraction: The phenomenon from relativity in which an object moving past an observer at high speed appears to be contracted in the dimension along the direction of its motion.

lepton: Along with quarks, 1 of the 2 families of particles that represent the current limit on our knowledge of the structure of matter at the smallest scales. The electron, muon, and tau particles, along with their antiparticles and their associated neutrinos, comprise the lepton family; only the electron is stable under the conditions present in the universe today.

lepton era: The era in the early universe between 10^{-4} and 1 second after the big bang in which the photons were energetic enough to make lepton-antilepton pairs (taus and muons in the beginning, only electrons near the end).

liquid: A state of matter in which the constituent atoms or molecules are touching but are free to slide over one another, allowing the matter to take the shape of the container in which it is held.

lysene: One of the 20 amino acids essential for life that contains 14 hydrogen atoms, 6 carbon atoms, 2 nitrogen atoms, and 2 oxygen atoms.

main sequence: The location on the H-R diagram in which roughly 90% of all stars are found. The sequence, running from cool (3000 K), low-luminosity (10^{-3} solar luminosities) stars to hot (50,000 K), high-luminosity (300,000 solar luminosities) stars is also a sequence in mass (0.08–80 solar masses) and lifetime (hundreds of billions to 3 million years). Stars are found on the main sequence during the longest-lasting phase of their lives, when they are in hydrostatic equilibrium and burning hydrogen to helium in their cores.

Michelson-Morley experiment: An experiment performed in the late 19th century by the 2 American physicists from which it takes its name, with the goal of detecting the presence of the aether through which electromagnetic waves moved. Its failure to detect any evidence for the aether led to the development of Einstein's theory of relativity.

microwaves: Electromagnetic waves with wavelengths from millimeters to centimeters in the radio part of the spectrum.

molecule: The combination of 2 or more identical or different atoms into a stable, single entity that will have chemical properties that differ from those of its constituents. The simplest molecules, such as the oxygen we breathe (O_2), consist of 2 identical atoms; complex molecules such as DNA—the template of life, which prescribes for our cells how to use oxygen as a fuel—contain tens of thousands of atoms of 5 different types (H, C, N, O, and P).

muon: The second generation of the lepton family, muons are unstable, decaying with a half-life of 10^{-6} seconds. Their production high in the atmosphere by cosmic rays and their detection on the ground provide a direct manifestation of relativistic effects in the real world.

neurons: Cells of the central nervous system consisting of a cell body, thousands of projections called dendrites, and a single long protrusion called the axon. These cells are specialized for detecting stimuli and transmitting them to other neurons in the system.

neutrino: The lightest of the subatomic particles with masses on the order of 1 millionth that of the electron. Neutrinos only interact via the weak nuclear force and, as such, pass through matter with ease. There are 3 flavors, corresponding to the 3 leptonic particles as well as to their antiparticles; neutrinos can transform themselves from one flavor to another in flight. They accompany all beta decays and can be used to peer directly at the nuclear furnace at the core of the Sun.

neutron: The neutral nuclear particle consisting of a bound triplet of 2 down quarks and 1 up quark. Free neutrons are unstable, decaying into a proton, an electron, and an antineutrino with a half-life of about 10 minutes. Bound in a nucleus, however, the neutron is stabilized and can live forever. Lacking a charge, the neutron does not contribute to the repulsive electromagnetic forces in the nucleus, but the remnants of the strong nuclear force binding the quarks together, which leak beyond the particle's boundary, helps add extra glue to keep the nucleus together. The right balance with protons is essential, however. With too many or too few neutrons, the nucleus will be unstable and subject to radioactive decay. The number of neutrons a nucleus contains determines the isotopic form of the element.

neutron activation: The process in which an item of interest is bombarded with a huge flux of neutrons, which transform some of the nuclei into radioactive isotopes. The subsequent decay of these isotopes reveals which elements are present in the object. This technique allows the nondestructive analysis of works of art and other materials.

neutron star: The end state of a massive star, in which gravitational collapse subsequent to the exhaustion of nuclear fuel crushes the electrons into the protons to form a mass composed primarily of neutrons. These stars have remarkable properties: a radius of just 10 kilometers, a density equal to that of nuclear matter (1 billion tons per teaspoonful), magnetic fields 10 trillion times that of Earth, and rotation speeds up to 600 rotations per second.

noble gas: One of the 6 elements in the right-hand column of the periodic table, which have complete outer shells of electrons and thus cannot participate in chemical reactions.

noble metal: Strictly speaking, a noble metal is an element in which the third energy sublevel is completely full, rendering it relatively nonreactive (i.e., copper, silver, and gold). More generally, it is any metal that does not react readily, especially with oxygen (and so includes rare metals such as iridium, platinum, and ruthenium).

octave: A change of a factor of 2 in the wavelength (or, equivalently, the frequency) of a wave. The human ear hears a range of 10 octaves, whereas the eye sees only 1 of the more than 50 octaves of electromagnetic radiation that arrive from space.

Ötzi: The well-preserved remains of a naturally mummified, Copper Age (3300 B.C.) resident of northern Italy discovered on the Austrian border in 1991. Isotopic analysis revealed striking details about his life and death.

periodic table: The arrangement in rows and columns of the elements that groups them by their chemical properties (now known to correspond to the consecutive filling of electron energy levels). First constructed by Dimitri Mendeleev.

photon: A packet of light energy. When an electron in an atom moves from one energy level to another, a photon can be either emitted or absorbed.

plasma: A state of matter in which the electrons are either wholly or partially separated from their atomic nuclei, requiring temperatures in excess of several thousand degrees. Most of the matter in the universe exists as a plasma.

potential energy: Stored energy, or the potential to do work. Water at the top of a waterfall, electrons at the pole of a battery, and a radioactive nucleus all have potential energy, which, when released, can do work.

proton: One of the 2 baryons prevalent in the universe today, the proton consists of 2 up quarks and 1 down quark bound together in a triplet by the strong nuclear force. It has a mass similar to the other baryon, the neutron, but has a charge of +1. The number of protons in a nucleus define which element an atom is; nuclei with 1 to 92 protons exist in nature.

quark: Along with leptons, 1 of the 2 families of particles that represent the current limit on our knowledge of the structure of matter at the smallest scales. There are 3 generations of quarks, each of which has 2 members—up and down, strange and charm, and top and bottom—plus their corresponding antiquarks, making a total of 12. Triplets of up and down quarks comprise the proton and neutron; pairs of quarks make particles called mesons. In the universe today, there are no "free" single quarks; all are bound up in pairs or triplets. The name, assigned by theoretical physicist Murray Gell-Mann, comes from a line in James Joyce's *Finnegan's Wake*.

quark era: The era in the early universe prior to 10^{-6} seconds after the big bang, in which the density of the universe was sufficiently high that the hadrons were overlapping and the quarks became free.

radiation: This term is used in different ways. Electromagnetic radiation includes radio waves; microwaves; infrared, visible, and ultraviolet light; X-rays; and gamma rays, spanning more than 50 octaves in wavelength. However, we also use the term to refer to high-energy subatomic particles such as electrons and positrons, protons, neutrons, and bare atomic nuclei that arrive from space or are the byproduct of radioactive decay.

radiation era: Using radiation in the first sense, the era of the universe from 1 second to 380,000 years after the big bang, in which the dominant mass-energy was in the form of photons (electromagnetic radiation). As the spacetime of the universe expanded, the wavelengths of these photons were stretched such that they lost energy, until at 380,000 years the balance shifted such that the dominant mass-energy was in particles (our current matter era).

radio waves: The portion of the electromagnetic spectrum in which the wavelengths are longer than 1 millimeter.

radioactivity: The process by which an atomic nucleus undergoes a spontaneous transformation to a more stable ratio of protons to neutrons and/or to a lower energy state. This transformation, while immune from external influences and occurring at a random time in any given nucleus, takes place in a collection of identical nuclei at a precisely determined average rate that can be used as an atomic timepiece. The interval over which half a sample decays (*see* **half-life**) ranges from fractions of a second to billions of years, depending on the type of nucleus.

red beds: Geological features representing the deposit of minerals containing large amounts of iron oxide, mostly formed after the banded iron formations.

red giant: The stage of the life of a star following exhaustion of hydrogen in the star's core. The core contracts such that first a shell of hydrogen surrounding the core burns, and then the helium left over from the first reaction becomes the fuel for a fusion to carbon. Since the energy given off is greater than in hydrogen fusion, the outer layers of the star expand and cool, creating a red giant star. The Sun will expand nearly 100-fold when it reaches this stage in about 6 billion years.

red shift: The shift toward the red end of the spectrum of light from distant objects that reveals the expansion of the universe. The effect is produced by the stretching of spacetime. As wavelengths of light pass through it, they are stretched, lowering their energies and shifting them toward the red end of the spectrum.

relativity: The model for spacetime developed by Albert Einstein in which lengths, times, masses, velocities, and the concept of simultaneity are all shown to be relative to the velocity of the observer.

right-handed molecule: *See* **left-handed molecule**.

secular equilibrium: The state in a radioactive decay chain when enough time has passed that all intermediate isotopes between the parent and ultimate (stable) daughter nucleus are present in amounts proportional to their half-lives.

snowball earth: The state, achieved several times in Earth's history, in which nearly all of the planet is covered by ice (land and oceans). The latest such episode was about 600 million years ago, just prior to the Cambrian explosion.

solid: The state of matter in which the atoms or molecules composing the material are touching and locked more or less rigidly in place by intermolecular forces.

space-time diagram: A 2-dimensional representation of spacetime in which the vertical axis represents time and the horizontal axis represents 1 dimension of space. The diagram makes the effects of relativity easy to visualize.

spacetime: The unification of space and time required by Einstein's theory of relativity as a consequence of the finite and immutable speed of light.

stomata: The microscopic openings on the underside of leaves through which carbon dioxide enters the leaf and oxygen and water vapor are exhaled.

stromatolites: Geological features that represent remnants of vast mats of cyanobacteria and thus the oldest fossilized life on Earth.

strong force: One of the 4 basic forces of nature, the strong force operates only on the scale of the atomic nucleus ($\sim 10^{-14}$ m) and vanishes beyond this scale. It applies to all particles that exhibit "color charge," a property ascribed to quarks. The strong force is responsible for holding the nucleus of an atom together against the repulsive electromagnetic force of the positively charged protons therein. Its messenger particle is the gluon.

sunspots: Dark patches on the surface of the Sun first catalogued by Galileo. They are now known to be magnetic storms erupting through the surface from the layer beneath. They are dark in visible light because the gas on the surface can be cooler, owing to the added support of strong magnetic fields; however, their associated ultraviolet and X-ray emission means the net effect is a slight (~0.1%) increase in solar output during the peak of the 11-year sunspot cycle. Longer-term variations in sunspot numbers may be linked to changes in Earth's climate on timescales of centuries to millennia.

tachyons: Theoretical particles that travel faster than the speed of light. There is no evidence whatsoever that tachyons actually exist, but relativity allows one to work out their properties.

temperature: The measure of the average kinetic energy of the particles in a substance. High temperature means the particles are moving rapidly, and low temperature means they are moving slowly.

time dilation: The relativistic effect that, for rapidly moving objects, time passes at a slower rate, often paraphrased as "moving clocks run slow." It is important to note, however, that it is time itself that is moving more slowly, not some malfunction of the clocks. The collision of muons produced high in the atmosphere with detectors on the ground demonstrates the reality of time dilation.

tree ring: The annual growth layer added to the outer circumference of a tree each year. The thickness and wood density as well as the isotopic composition of the rings provide clues to climate, while simply counting the rings provides a precise calendar. Using living and dead trees, a continuous record extending nearly 12,000 years has been accumulated for the Northern Hemisphere.

ultraviolet: The portion of the electromagnetic spectrum with wavelengths just short of visible light.

watt: A measure of power, 1 watt is the expenditure of 1 joule of energy in 1 second.

wave: An undulation that transmits energy from one place to another without any physical matter traveling through the space separating the 2 locations. The disturbance pattern can travel in the same direction of the wave (as in sound) or perpendicular to it (as in water waves and light).

wavelength: The distance between 2 crests of a wave. The wavelengths of visible light stretch from 350 nanometers to 700 nanometers, or 1 octave, while the electromagnetic spectrum stretches from radio waves, with wavelengths of thousands of kilometers, to gamma rays, with wavelengths of 10^{-21} m.

weak force: One of the 4 fundamental forces of nature, the weak force, like the strong force, only operates on the scale of the nucleus of an atom ($\sim 10^{-14}$ m). The weak force mediates radioactive decay and often involves the neutrino. Its carrier particles are the intermediate vector bosons, the $W^{+/-}$ and the Z^0.

white dwarf: The end state for all stars less massive than 8 solar masses. White dwarfs have diameters approximately equal to that of the Earth, or 1% the size of their parent stars, and have densities of roughly 1 ton per teaspoonful. Since their nuclear reactions have ceased, they have no source of internal energy, so they simply radiate away the heat they have when formed, like a dying ember in a fire.

X-rays: The portion of the electromagnetic spectrum with wavelengths ranging from about 10 nanometers to 0.01 nanometers. Since their wavelengths are comparable to or smaller than individual atoms, they can easily penetrate even solid materials. The innermost electrons of atoms ranging from carbon to uranium are bound with high energies such that they emit X-rays when they fall into place.

Biographical Notes

Alvarez, Walter (b. 1940): Alvarez was born in California and received his Ph.D. from Princeton in 1967. In the late 1970s, he discovered that the rock layer separating the geologic eras known as the Cretaceous and Tertiary—the layer dividing sediments bearing dinosaur fossils from those in which all dinosaurs had vanished—contains an extraordinarily high concentration of the rare element iridium. In 1980, he published a paper attributing this elemental anomaly to the impact of a meteorite and speculating that the mass extinction that occurred at this boundary in the geologic record was caused by the consequences of this impact. While initially ridiculed, this hypothesis has come to be almost universally accepted; the crater produced by the impact has been found, and numerous other strands of evidence support this scenario.

Democritus (460–370 B.C.): Democritus was the dominant pre-Socratic philosopher of the ancient world. He spent his life in travel and research, developing a philosophical outlook that is decidedly modern and scientific in comparison with that of Socrates, Plato, and Aristotle—he sought causes in the material world rather than pursuing the Aristotelian emphasis on purpose. In our context, he is the first (along with his teacher Leucippus) to postulate the concept of the atom as the smallest indivisible unit of matter.

Douglass, Andrew Ellicott (1867–1962): Douglass was born in Vermont and, at the age of 27, became an assistant to Percival Lowell at the Lowell Observatory in Arizona. He immediately began the study of patterns in tree-ring growth, picking up on an idea first broached by Leonardo da Vinci more than 4 centuries earlier—that one could read the "history of past seasons" in the varying thicknesses of tree rings. He endeavored to find a correlation between the sunspot cycle and climate, a relationship that remains somewhat controversial today. In the process he established the science of dendrochronology and founded the Tree-Ring Laboratory at the University of Arizona, which remains a dominant center of activity in the field today.

Einstein, Albert (1879–1955): Einstein was born in what is now Germany and was educated there and in Switzerland. Of his many contributions to physics, most relevant here is his theory of relativity, first published in 1905. Built on the work of other physicists over the preceding decades, relativity dissolves the separate natures of space and time and of matter and energy, in the latter case through the famous relation $E = mc^2$. In addition, however, in the same year, Einstein published a seminal paper on Brownian motion that played a major role in establishing the size and mass scales of atoms.

Galilei, Galileo (1564–1642): Galileo was born in Tuscany and educated at Pisa. His foundational influence in astronomy and physics is difficult to overestimate. His telescopic observations demolished the Earth-centered cosmology of the ancients, confirmed in detail Copernicus' Sun-centered solar system, and recognized the Sun was but one of tens of thousands of stars that, he speculated, hosted planetary systems of their own (an idea only confirmed in the past 15 years). His observations of the heavens and his experiments in his laboratory laid the groundwork for Newtonian mechanics. His recognition of the importance of mathematics in describing the physical world forms the basis for the development of all of modern science.

Hubble, Edwin (1889–1953): Hubble was born in Missouri and, after becoming one of the first Rhodes Scholars at Oxford and serving in World War I, obtained his Ph.D. in 1917 from the University of Chicago. He subsequently moved to the Carnegie Observatories in Pasadena, California, where he remained for the rest of his life. His discovery that the spiral nebulae could be resolved into individual stars radically expanded our view of the universe as extending far beyond the confines of the Milky Way galaxy. He developed techniques to measure the distances to these other galaxies and, coupling these with Vesto Slipher's measurements of velocities, plotted his famous Hubble diagram in 1929, showing that the rate of recession was proportional to distance. The slope of the line linking these 2 quantities, now called the Hubble constant, sets the scale of the universe.

Kelvin, Lord (a.k.a. **William Thompson**; 1824–1907): Kelvin was an Irish-born, Scotland- and Cambridge-educated physicist who left his mark on many areas of physics over his long career. His relevance to this series is 2-fold. First, his invention of the Kelvin temperature scale recognized the physical basis of temperature as the motion of atoms and molecules by establishing the zero point at the temperature where atomic motion ceases. Second, throughout the middle and late 19th century he led the physicists' denial of the long time scales required by geologists and evolutionary biologists, insisting that the solar system could be no more than 100 million years old. He did leave himself an out, however, with his famous line "… unless sources now unknown to us are prepared in the great storehouse of creation."

Lavoisier, Antoine (1743–1794): Lavoisier was born and educated in Paris. His numerous accomplishments include the first promulgation of the law of the conservation of mass in chemical reactions, the recognition that air was a mixture of gases, a major role in the construction of the metric system, and the discovery and naming of many elements, including oxygen. He is sometimes referred to as the father of modern chemistry. He was beheaded during the French Revolution.

Libby, Willard (1908–1980): Libby was born in Colorado and obtained his Ph.D. from Berkeley in 1933. In 1949, while on the chemistry faculty of the University of Chicago, Libby invented the carbon-14 dating technique that has become the most useful method of radiometric dating for artifacts less than 60,000 years old and has revolutionized the fields of history and archeology. He was awarded the Nobel Prize for this achievement in 1960.

Mendeleev, Dmitri (1834–1907): Mendeleev was born and educated in Russia, obtaining his Doctorate from the University of St. Petersburg in 1865. He made many contributions of note to the fields of physics, chemistry, and geology, but his principal claim to fame is his creation of the periodic table of the elements. The scientific triumph of this work, first presented in 1869, was that, from the patterns of weights and chemical behaviors of the elements, he predicted that several elements with specific properties were yet to be discovered. It took more than a quarter of a century to find germanium, for example, but its properties fit precisely those predicted by Mendeleev in his original presentation of the table.

Milankovich, Milutin (1879–1958): Milankovitch was born in what is now Croatia and received his Ph.D. at the Vienna Institute of Technology in 1904. In 1909 he obtained a position at the University of Belgrade, which he held for the remainder of his life. During the First World War, while imprisoned by the Austro-Hungarian Empire, he carried out an elaborate series of calculations that quantified the subtle changes in the Earth's orbit and orientation on time scales of tens to hundreds of thousands of years caused by the gravitational perturbations of the Sun, the Moon, and other planets. Although he published this work in 1920, it was not until the 1960s that it was rediscovered and shown that these changes were responsible for the quasi-periodic ice ages of the past several million years.

Lowell, Percival (1855–1916): Lowell was born into the "Boston Brahmin" Lowell family and was educated at Harvard. Following a period of travel in the Far East that led to several books, he turned to the study of astronomy. Using his personal wealth, he created the first high-altitude, remote astronomical observatory at Flagstaff, Arizona, with the express intention of studying the so-called canals on the surface of Mars. He developed an elaborate story surrounding his observations: A dying civilization in the throes of a millennial drought was desperately building a vast network of canals to bring water from the poles to the parched equator. While his work led to popular acclaim, the scientific community remained highly skeptical. His observatory did, however, produce 2 other characters in our story, Vesto Slipher and A. E. Douglass.

Rabi, Isidor Isaac (1898–1988): Rabi was born in what is now Poland but moved to the United States a year later and received his Ph.D. from Columbia in 1927. Remaining at Columbia for more than 60 years, he became one of the leading figures of 20th-century physics, as well as playing a central role in public policy during the middle decades of the century. Following his receipt of a Nobel Prize in 1944, Rabi chaired the Columbia Physics Department for 5 years, during which an astonishing 13 future Nobel Prize–winning faculty and students were mentored. In our context, he is responsible for creating the method of molecular beam magnetic resonance, which paved the way for the invention of the atomic clock.

Russell, Henry Norris (1877–1957): Russell was born in New York and received his Ph.D. from Princeton in 1899. He returned to Princeton in 1905 to join the faculty and remained there until his death. He played a major role in the development of stellar astrophysics in the first half of the 20th century. His plot of stellar surface temperature versus total energy output (luminosity), named for him and for its codiscoverer Ejnar Hertzprung, is called the Hertzsprung-Russell (or H-R) diagram and is the key to unlocking the secrets of stellar evolution.

Slipher, Vesto (1875–1969): Slipher was born in Indiana and received his Ph.D. in astronomy in 1909. He promptly moved to the Lowell Observatory in Flagstaff, Arizona, where he remained for his entire career, serving as director from 1915 to 1952. His principal contribution was the determination of the recession velocities of several dozen nearby galaxies (or "spiral nebulae," as they were called before being recognized as galaxies of stars separate from the Milky Way). All but a few of the 46 objects he measured were receding from Earth, such that their light was "red shifted," or moved toward the red end of the electromagnetic spectrum by the Doppler effect. Edwin Hubble used Slipher's data in producing his famous plot that showed that the distance of a galaxy from Earth was proportional to its recession velocity, thus laying the observation basis for the expanding universe.

Gell-Mann, Murray. *The Quark and the Jaguar: Adventures in the Simple and Complex.* New York: Holt Paperbacks, 1995. A fascinating book that touches on several of the topics in these lectures, written by the 1969 Nobel laureate in Physics who invented quarks.

Greene, Brian. *The Elegant Universe: Superstrings, Hidden Dimensions, and the Quest for the Ultimate Theory.* New York: Vintage, 1999. Written by my string-theorist colleague Professor Greene, this weighty book is viewed by many as an excellent introduction to what we know about particles on the one hand and the universe on the other, and how modern scientists are trying to come up with a theory that will unite that knowledge.

Hale, John, Jan Heinemeier, Lynne Lancaster, Alf Lindroos, and Åsa Ringbom. "Dating Ancient Mortar." *American Scientist* 91 (March–April 2003): 130–137. This paper, an excellent example of how scholars from different areas can work together to solve a complex problem, forms the basis of Lecture Seven; more details and illustrations can be found here. Also available online (http://www.americanscientist.org/issues/feature/2003/2/dating-ancient-mortar/1).

Harrison, Edward. *Cosmology: The Science of the Universe.* 2nd ed. Cambridge: Cambridge University Press, 2000. In my view, this is the most literate and elegant book on relativity and cosmology ever written.

Helfand, David J. *Scientific Habits of Mind.* http://www.fos-online.org/habitsofmind/index.html. This is a free electronic book I have written to accompany the Columbia Core Curriculum course Frontiers of Science; it covers many basic quantitative reasoning skills and, through personal anecdotes, introduces the ways they can be utilized in everyday life.

Hitch, Charles J. "Dendrochronology and Serendipity." *American Scientist* 70 (May–June 1982): 300–305. This paper inspired Lecture Ten; details and illustrations can be found here.

Jerome, Kate Boehm. *Atomic Universe: The Quest to Discover Radioactivity.* Washington DC: National Geographic, 2006. A brief history of how we came to understand the strange and energetic "emanations" some atoms produce.

Kirshner, Robert P. *The Extravagant Universe: Exploding Stars, Dark Energy and the Accelerating Cosmos.* Princeton, NJ: Princeton University Press, 2004. Professor Kirshner, former chair of the Department of Astronomy at Harvard, is an irreverent and humorous sort who describes the latest findings in cosmology (particularly the accelerating universe) with verve and style.

Kunzig, Robert, and Wallace S. Broecker. *Fixing Climate: The Story of Climate Science—And How to Stop Global Warming*. New York: Hill and Wang, 2008. Kunzig is an excellent science writer; the book records the story of how past changes in Earth's climate hold the key to understanding future changes and what to do about it, as told by the National Medal of Science winner and the Grandfather of Climate Change, my colleague Professor Broecker.

Lambert, Joseph B. *Traces Of The Past: Unraveling The Secrets Of Archaeology Through Chemistry*. Reading, MA: Helix Books, 1998. A leading archeological chemist shows how molecules contribute to our understanding of prehistoric materials—stone, soil, pottery, color, glass, organics (including plaster and food), and metals.

Lane, Nick. *Power, Sex, and Suicide: Mitochondria and the Meaning of Life*. New York: Oxford University Press, 2005. A brilliant book about how mitochondria invaded single-celled organisms 2 billion years ago and have shaped the evolution of life ever since.

Lichtenberg, Don. *The Universe and the Atom*. Hackensack, NJ: World Scientific, 2007. An account that covers all the scales of this course and draws interesting connections among them, starting from the world as we sense it and showing how limited our senses are.

McBride, Neil, and Iain Gilmour, eds. *An Introduction to the Solar System*. Cambridge: Cambridge University Press, 2004. Two chapters by Ian Wright at the end of the book (chapters 7 and 8) discuss solar system evolution and meteorites.

McGraw, Donald. "Andrew Ellicott Douglass and the Big Trees." *American Scientist* 88 (September–October 2000): 440–447. A historical recounting of the founding of a new science: dendrochronology.

Morrison, Philip and Phylis Morrison. *Powers of Ten: A Book about the Relative Size of Things in the Universe and the Effect of Adding Another Zero*. Redding, CT: Scientific American Library, 2000. This book accompanies a classic short film that takes a journey through the size scales of the universe.

Nash, Stephen E. *Time, Trees, and Prehistory: Tree-Ring Dating and the Development of North American Archeology, 1914–1950*. Salt Lake City: University of Utah Press, 1999. A description of the major changes that the development of dendrochronology brought to our understanding of American archeology prior to the introduction of carbon-14 dating.

Paulos, John Allen. *Innumeracy: Mathematical Illiteracy and Its Consequences*. New York: Hill and Wang, 2001. A classic short book about the pitfalls of innumeracy and some cures for this affliction; humorous and inventive.

Rees, Lord Martin. *Our Cosmic Habitat*. Princeton, NJ: Princeton University Press, 2001. While I do not subscribe to all of the author's philosophical leanings, he is a superb astrophysicist and a superb writer for the public—a rare combination of talents.

Rhodes, Richard. *The Making of the Atomic Bomb*. New York: Simon and Schuster, 1986. Pulitzer Prize–winning tome on the creation of first atomic bomb. The first 200 pages provide an excellent introduction to the development of our understanding of the atom in highly literate and accurate prose.

Rucker, Rudy. *The Fourth Dimension*. New York: Houghton Mifflin, 1984. A slightly wacky and totally inspired book about thinking in higher dimensions. It even includes a demonstration!

Spindler, Konrad. *The Man in the Ice: The Discovery of a 5,000-Year-Old Body Reveals the Secrets of the Stone Age*. New York: Three Rivers Press, 1996. A firsthand account by the archeologist who studied the remains of Ötzi the Iceman, found in 1991 near the Austro-Italian border.

Stewart, Ian. *Flatterland: Like Flatland Only More So*. Cambridge, MA: Perseus Publishing, 2001. The modern accompaniment to *Flatland* by an accomplished mathematician and science writer, it introduces the concepts of non-Euclidean geometries essential for the study of cosmology.

Taft, W. Stanley, James W. Mayer, Richard Newman, Dusan Stulik, and Peter Kuniholm. *The Science of Paintings*. New York: Springer, 2001. The text covers such topics as paints, binders, optics, and detection of forgeries (chapter 8), while lengthy appendices cover additional scientific topics such as nuclear reactions and neutron-activation analysis (Appendix G), radiocarbon dating in art research (Appendix K), and dendrochronology of panel paintings (Appendix L).

Taylor, Edwin, and John Archibald Wheeler. *Spacetime Physics*. 2nd ed. New York: W. H. Freeman, 1992. A classic textbook for understanding special relativity. It was originally published in 1966 but has been updated to include a final chapter on the general theory of relativity. It includes many problems with which the reader can test his or her understanding of the topics covered.

Thomas, Lewis. *The Lives of a Cell*. New York: Viking Press, 1974. In my view, no one has ever written more beautiful prose on science than Thomas; all of his essay collections contain incisive insights into our modern understanding of the universe delivered in language that is at once precise and poetic.

Van der Merwe, Nikolaas J. "Carbon Isotopes, Photosynthesis and Archaeology." *American Scientist* 70 (November–December 1982): 596–606. It was after reading this article that the idea for this course began to take shape, since it provides such an accessible account of a beautiful example of an atomic reconstruction of otherwise inaccessible history.

Walker, Gabrielle. *An Ocean of Air*. New York: Houghton Mifflin Harcourt, 2007. A history not of the atmosphere itself as much as of our discoveries about the ocean of air under which we live.

Woolfson, Michael. *The Formation of the Solar System: Theories*. London: Imperial College Press, 2007. A British physicist gives rather exhaustive consideration of various theories of the solar system, with consideration of isotopes in meteorites toward the end of the book.